Observing the Universe

NEW SCIENTIST GUIDES
Series Editor: *Colin Tudge*

The Understanding of Animals
Edited by *Georgina Ferry*

Observing the Universe
Edited by *Nigel Henbest*

Grow Your Own Energy
Edited by *Mike Cross*

Building the Universe
Edited by *Christine Sutton*

The Making of the Earth
Edited by *Richard Fifield*

A new **scientist** GUIDE

OBSERVING THE UNIVERSE

Edited by
NIGEL HENBEST

Basil Blackwell & New Scientist

© Articles and editorial, IPC Magazines Ltd.
Volume rights, Basil Blackwell Limited.

First published in book form in 1984 by
Basil Blackwell Limited.
108 Cowley Road, Oxford OX4 1JF.

All rights reserved. Except for the quotation of short passages for the purpose of criticism and review, no part of this publication may be reproduced, stored in a retrieval system, or transmitted, in any form or by any means, electronic, mechanical, photocopying, recording or otherwise, without the prior written permission of the copyright holders.

Except in the United States of America, this book is sold subject to the condition that it shall not, by way of trade or otherwise be lent, re-sold, hired out, or otherwise circulated without the publisher's prior consent in any form of binding or cover other than that in which it is published and without a similar condition including this condition being imposed on the subsequent purchaser.

British Library Cataloguing in Publication Data

Observing the universe. – (New scientist guides)
1. Cosmology
I. Henbest, Nigel
523.1 QB981

ISBN 0-85520-727-2
ISBN 0-85520-726-4 Pbk

Typeset by Oxford Verbatim Limited
Printed in Great Britain by
Bell and Bain Ltd., Glasgow

The 11¾ inch Mertz refractor erected at Greenwich in 1859

Contents

List of Illustrations

Frontispiece, C. Sutton; p. 3 Hale Observatories: p. 5 David Malin: p. 8 South African Astronomical Society: p. 20 William D. Pence: p. 45 British Museum: p. 50 Sterrewacht, Leiden; Huygens Laboratory: p. 53 Space Frontiers Limited: p. 57 Appleby-Boningham Steel Company: p. 59 Central Press Photos Ltd.: p. 63 Hamburger Aerolloyd; Reg.-Präs Dusseldorf: p. 65 Max Planck Institut für Radio-Astronomie: p. 67 C. S. I. R. O. Radiophysics Division Photo Lab.: p. 73 BICC Ltd.: p. 74 David Malin: p. 79 Jerry Mason: p. 90 NASA: p. 92 Leicester University, Science Photo. Library: p. 107 NASA: p. 119 Perkin-Elmer Corporation: p. 123 ESA: p. 134 Anglo-Australian Observatory: p. 136 NASA: p. 139 Royal Observatory, Edinburgh: p. 144 W. J. Robertson LIIP; Royal Observatory, Edinburgh: p. 155 NASA: p. 160 NRAD: p. 164 Rutherford Appleton Laboratory: p. 173 Photolabs; Royal Observatory, Edinburgh: p. 183 Gary Higbee: p. 199 Royal Greenwich Observatory: p. 200 Anglo-Australian Observatory: p. 202 Associated Press: p. 205 Australian Information Services: p. 214 SERC: p. 233 D. Calvert; R. Worth; Royal Greenwich Observatory: p. 238 D. Calvert; R. Worth; Royal Greenwich Observatory: p. 242 N. Parsons; p. 245 Smithsonian Institute: p. 254 Kitt Peak National Observatory: p. 275 NASA: p. 278 Perkin-Elmer Corporation.

Contributors

NIGEL CALDER wrote about Jodrell Bank while a staff writer on *New Scientist*; he was later editor (1962–66). His science television programmes and accompanying books include *Violent Universe*, *Einstein's Universe*, *The Restless Earth* and *The Comet is Coming* (BBC).

DR MALCOLM COE is a Lecturer in the gamma-ray astronomy group in the Department of Physics, University of Southampton.

PROFESSOR LEN CULHANE is Deputy Director of the Mullard Space Science Laboratory, the X-ray astronomy research group of University College, London.

DR JON DARIUS wrote about the IUE satellite while at the Department of Physics, University College, London. He subsequently spent two years working for ESA as resident astronomer on IUE at the satellite tracking station near Madrid and is now at the Science Museum, London.

PROFESSOR RONALD DREVER is at the Department of Natural Philosophy, University of Glasgow.

PROFESSOR ANDY FABIAN wrote on the Copernicus satellite while at the Mullard Space Science Laboratory, University College, London. He is now at the Institute of Astronomy, Cambridge.

PROFESSOR PETER FELLGETT is professor of cybernetics and instrument physics, University of Reading.

NIGEL HENBEST is astronomy consultant to *New Scientist*. His books include *The Restless Universe* (with Heather Couper; George Philip, 1982) and *The New Astronomy* (with Michael Marten; Cambridge University Press, 1983).

ROS HERMAN is a science editor on *New Scientist*, with an interest in science policy.

DR MICHAEL HILLAS is at the Department of Physics, University of Leeds.

DR KEITH HINDLEY is a freelance writer, specialising in astronomy and environmental subjects.

DR JAMES HOUGH is at the Department of Natural Philosophy, University of Glasgow.

PROFESSOR SIR FRED HOYLE is an honorary research professor at Manchester University and University College, Cardiff. At the beginning of the period reviewed here (1956) he was lecturer in mathematics at the University of Cambridge, and he was later head of the Institute of Theoretical Astronomy at Cambridge. His popular books include *Lifecloud* (with N. C. Wickramasinghe; Dent, 1978).

DR EDWARD KIBBLEWHITE is Assistant Director of Research at the Institute of Astronomy, Cambridge.

DR ED KRUPP is director of the Griffith Observatory, Los Angeles, and is author of *In Search of Ancient Astronomies* (Penguin, 1980)

MICHAEL MAUNDER is at the Department of Industry, and has developed photographic processing for astronomical use.

PROFESSOR BERNARD Y. MILLS is at the Department of Astrophysics, University of Sydney.

ANDREW MURRAY is at the Royal Greenwich Observatory, working on the Hipparcos satellite.

PROFESSOR KEN POUNDS heads the X-ray astronomy group in the Department of Physics, University of Leicester.

PROFESSOR RODERICK REDMAN, who died in 1975, wrote on the Anglo-Australian Telescope while Plumian Professor at the University of Cambridge.

PROFESSOR SIR MARTIN RYLE is emeritus professor at the Cavendish Laboratory, University of Cambridge. He was Astronomer Royal from 1972 to 1982, and Nobel Laureate in physics in 1974.

PROFESSOR F. GRAHAM SMITH wrote on the birth of radio astronomy while at the Cavendish Laboratory, Cambridge. He was director of the Royal Greenwich Observatory (1976–81), and is now director of the Nuffield Radio Astronomy Laboratories, Jodrell Bank, and is Astronomer Royal.

DR F. RICHARD STEPHENSON wrote on Chinese and Babylonian astronomy while at the University of Liverpool. He is now at the Department of Physics, University of Durham. His books include *The Historical Supernovae* (with David H. Clark, Pergamon, 1977).

DR PETER STUBBS wrote on the Effelsberg radio telescope while deputy editor of *New Scientist.*

DR CHRISTINE SUTTON is a freelance science writer, who wrote on the IRAS satellite while physical science editor of *New Scientist.*

DR ALAN WATSON is reader in particle cosmic physics in the Department of Physics at Leeds University and has worked at the Haverah Park Array since 1964.

CAPTAIN SIMON WORDEN is an astronomer who wrote on speckle interferometry while working at the US Air Force Geophysics Laboratory and the Sacramento Peak Observatory, New Mexico. He is now at the USAF HQ Space Division, Los Angeles.

A Note on Units

Wavelength

Wavelengths shorter than a millimetre are given in micrometres or in nanometres.

1 micrometre $= 10^{-6}$ metre
1 nanometre (nm) $= 10^{-9}$ metre

The shortest waves, gamma rays, are usually quoted in terms of the energy of their photons, which is inversely proportional to the wavelength. The basic energy unit is the electronvolt (eV).

1 eV $= 1.6 \times 10^{-19}$ joule
1 keV $= 10^{3}$ eV
1 MeV $= 10^{6}$ eV
1 GeV $= 10^{9}$ eV
wavelength (in nm) = 1.24/energy (in keV)

Cosmic ray energies are also given in eV.

Wavelength ranges

Radio waves	longer than 1 millimetre
millimetre waves	1–3 millimetres
Infrared	0.7–1000 micrometres
submillimetre waves	300–1000 micrometres
Optical	390–700 nanometres
Ultraviolet	10–390 nanometres
extreme ultraviolet	10–91 nanometres
X-ray	0.01–10 nanometres
Gamma ray	shorter than 0.01 nanometres

Angular size

1 degree = 60 arcminutes
1 arcminute = 60 arcseconds

1 degree is roughly twice the apparent size of the Moon.
1 arcminute is about the finest detail the human eye can resolve; a little larger than the apparent size of Venus or Jupiter.
1 arcsecond is the limit of resolution for ground-based optical telescopes, set by turbulence in the air above.

Magnitudes

Optical astronomers quote apparent brightnesses in the archaic system of magnitudes. Fainter objects have larger magnitudes; and a difference of 5 in magnitude corresponds to a ratio of 100 in brightness.

brightest star, Sirius	magnitude −1.4
faintest star visible to eye	6.5
faintest object visible to ground-based telescope	23
faintest object visible to Space Telescope	27

Temperature

Temperatures are given in degrees Kelvin (absolute), which starts from absolute zero: 0 K is −273°C, 273 K is 0°C, and so on.

Distance

Distances are given in light years (although astronomers also use another unit, the parsec).

1 light year = 9.46×10^{12} kilometres

PART ONE

Telescoping the Cosmos

Astronomy claims to be the oldest of the sciences, but it is in no way an elderly science. Astrophysicists and cosmologists are today studying many of the most exciting, controversial and fundamental fields of research, for the Universe is a laboratory where we find conditions that we can never hope to achieve on Earth. Diffuse interstellar clouds, for example, demonstrate chemical reactions that would require immense lengths of time to come to equilibrium in a terrestrial laboratory; and the condensed neutron stars, a thousand million million times denser than water, indicate what happens when matter is compressed far more than anything we can produce on our planet. The Universe provides physicists, *gratis*, the energy concentration of quasars and the irresistible gravitational pull of black holes. And astronomical observations alone can answer the fundamental questions about the Universe: what is it composed of; how old is it; how did it all begin?

But as a laboratory the Universe has some unique disadvantages. We cannot experiment in the normal sense, turning one knob and then another to see how the object of investigation responds. It is not possible to alter conditions in the Universe at large. We must passively observe what is happening, and make deductions from the observations.

As observational scientists, astronomers depend very much on the tools they use to observe – tools we can describe under the catch-all title "telescopes". With merely our own senses – the unaided power of the human eye – we might still be arguing whether the Earth moves around the Sun, and certainly the question of whether the stars are distant suns or specks of light on a black dome could be decided only by one's philosophical preference. "Astronomy" as a science could not exist without "telescopes".

There have been two great revolutions in astronomy. The first accompanied the surge of the Renaissance. The transition from medieval to modern thought brought with it the bold proposal that the Earth circles the Sun, a theory propounded by Nicolaus Copernicus in 1543, formulated mathematically by Johannes Kepler half a century later, and given a physical basis in Isaac Newton's theory of gravitation, published in 1687. But equally important in this revolution was the development of the optical telescope. Galileo's observations, begun in 1609, had the greatest impact. His discovery of the moons of Jupiter and the phases of Venus were unambiguous evidence in favour of the Sun-centred Universe. He also found that the Milky Way is a great amalgam of individual faint stars.

The legacy of the first astronomical revolution has lasted until surprisingly recently. For three and a half centuries after Galileo's pioneering observations, the chief concern of astronomers continued to be the study of planets and stars, using optical telescopes planted firmly on the ground.

The second astronomical revolution has come in our own times. It has sprung almost entirely from the use of new kinds of telescope – instruments to detect the long-wavelength radio and infrared radiations; to record the short-wavelength ultraviolet, X-rays and gamma rays; to trap high-energy cosmic ray particles and to search for gravitational waves, ripples in the background of space-time. Much of this research is only possible with telescopes above the Earth's atmosphere, and the astronomical revolution has advanced with the burgeoning "space age". Before this new era, astronomers knew only of radiation sources which emit light: nebulae, stars and the distant "star-islands" of the galaxies. Because these objects should emit little radiation other than light, astronomers generally saw little point – if they thought about it at all – in investigating the Universe at wavelengths where they expected to "see" nothing.

The new telescopes have, however, revealed that the Universe is full of objects that produce other radiations – and often very little light. The brightest infrared sources are luminous embryo stars deep within dark dust clouds which hide them from optical astronomers. The most brilliant gamma ray star is the Vela pulsar, a small neutron star emitting so little light that it was detected optically only as recently as 1977, with the world's most sensitive light detector attached to one of the largest optical telescopes. Other non-optical telescopes have picked up radio waves generated in magnetic clouds millions of light years long, stretching out from active galaxies, and X-rays from huge regions of hot gas in clusters of galaxies. In both these cases the regions of sky appear to optical telescopes merely as "empty space".

Optical astronomy has by no means been superseded by these other telescopes. Light from all kinds of objects carries interesting information, especially in its spectral lines, and optical telescopes are relatively simple and cheap to build. The momentum of the new revolution has swept up optical astronomers too. No longer are optical telescopes pointed only at stars and planets. Today they spend more time engaged at the forefront of astrophysics and cosmology, investigating distant galaxies and quasars – explosions at the centres of remote galaxies which outshine their parent galaxies a hundred times over. Advances in computing, electronics and

The core of this spiral galaxy, NGC 4151, is a "mini-quasar" – a compact but extremely powerful source of radiation of all lengths from gamma ray to radio waves. Recent observations with the International Ultraviolet Explorer satellite indicate that the mini-quasar is a disc of hot gas surrounding a black hole that weighs as much as 100 million suns

analysis, invented of necessity for radio telescopes and space astronomy, have now been harnessed by optical astronomers to improve the efficiency of optical telescopes, light detectors and the analysis of photographs and television images taken with modern optical telescopes.

The new discoveries have advanced the theoretical side of astrophysics and cosmology to areas undreamed of earlier this century. Before the revolution in telescopes, one of the main problems was seen to be the production of elements in stars; and the controversy over the origin of the Universe – big bang or steady state? – was a philosophical question, with no observational evidence either way. Now ordinary stars have become a mundane part of astrophysics. Most theorists have turned to the more esoteric fields opened by the new observations: interstellar chemistry, the solid-state properties of neutron matter in bulk, the natural particle accelerators that produce cosmic rays in our Galaxy and the huge magnetic clouds of radio galaxies, and the behaviour of matter near a black hole, to name a very few. On the cosmological side, the discovery of the microwave background radiation has convinced most astronomers that the Universe did indeed begin in a big bang, which has left this all-pervasive background of radio waves from its cooled-down gases. Theorists are now calculating what happened in the first fraction of a second after the big bang, back to the time when the Universe was only 10^{-35} seconds old.

This new understanding of the Universe has also, paradoxically enough, reawakened interest in the most ancient observations of the sky. Historical records stretch back for centuries before the first telescopic observations, in the case of the Chinese to the 2nd century BC. The human eye is a very useful "telescope" itself, because it possesses one unique property: an extremely long time base over which it has been recording the sky. For example, the gradual slowing down of the Earth's rotation shows up clearly in ancient records of solar eclipses, because over the centuries the cumulative change in rotation has displaced the Moon's shadow thousands of kilometres from where the eclipse "should" have been seen. Naked-eye monitoring of the sky has also recorded events that occur so rarely that modern detectors have yet to see one. The prime example is the appearance of supernovae in our Galaxy. These exploding stars are so rare that none has been recorded for over three centuries. Yet historical records, particularly those of the Chinese, tell us the positions and brightness of half a dozen such "guest stars" over the past 2000 years, and the accounts have been invaluable in guiding radio and X-ray astronomers to the wreckage of nearby supernova explosions.

The "new astronomy" all seems perfectly natural to today's young research students, but it has occurred only very recently – during the lifetime of many of us. It is difficult, however, to pin down the start of the revolution to one particular event or year. American engineer Karl Jansky accidentally discovered the first cosmic radio waves in 1932, but it took 20 years for astronomers to realise the importance of their radio colleagues' work. Perhaps the revolution dates from the identification of Cygnus A in 1951, when astronomers found themselves

The first identified quasar, 3C 273, was located in 1962. Already known as a radio source, its position was calculated accurately when the Moon passed in front of it and blocked off its radiation. It turned out to be a star-like object – blurred into a large image here by the Earth's atmosphere – with a faint jet, extending down to the lower right

faced with a galaxy pouring out as much energy in the form of radio waves as in light from its stars. Another milestone was the exciting period 1962–63, when the first quasar (radio designation 3C 273) was identified, and X-ray astronomy was born with the discovery of the first cosmic X-ray source Scorpius X-1. The quasar lies 2000 million light years away, while Scorpius X-1 is within our Galaxy, but both indicated that the Universe contains regions where energy is intensely concentrated.

Whatever the exact date we choose, these first successes of radio astronomy in the 1950s and X-ray astronomy in the 1960s led directly to the great astronomical revolution of our times, a revolution that has left textbooks of 30 or 40 years ago almost as outdated as those of the last century. *New Scientist* was founded in the early years of the revolution, in 1956, when radio astronomy was one of the exciting new branches of science that was catching the public's imagination. Over the years the magazine has kept readers informed of the new techniques of observation, as well as the results. This volume is a compilation of articles describing the innovative telescopes that have led to the greatest breakthrough in understanding the Universe since the Renaissance – and the even more imaginative telescopes now being designed and built.

1

The big bang in astronomy

FRED HOYLE

19 November 1981

Sir Fred Hoyle has been intimately involved in the "second revolution" in astronomy, taken here as the 25 years from 1956 to the original publication of this article. His personal account of the development of theory and of the crucial new observations leads him to review critically the generally accepted big bang theory of the Universe's beginning – and pose questions which orthodox astronomy has yet to ask, let alone answer.

During the spring and early summer of 1956, Willy Fowler, Margaret and Geoffrey Burbidge and I laid the foundation for the paper which became known later as B^2FH. The moment was astronomically ripe from two points of view. Through the early 1950s astronomers had begun to suspect that stars vary in their chemical compositions. Indeed the detection of such variations was regarded in the mid-1950s as the big problem of the day. It was also ripe for widening the breach created in the late 1930s by Hans Bethe, the breach for the entry of nuclear physics into astronomy. How far this breach was to go, none of us working on B^2FH could conceive. I attended in July [1981] a two-week summer school, organised by Martin Rees in Cambridge, given over to the study of supernovae, one of the sections of B^2FH. The experience for me was uncanny and a little sad. Uncanny because thoughts which at one time seemed personal (personal property almost!) had become the small change of everyday life to a later generation, and uncanny because of the tremendous technical development of the subject. I felt as Thomas Edison might have felt if he were to have listened to modern hi-fi equipment.

Sad, too, for at that two-week meeting I itched to be in there fighting it out with the boys of the new generation, just as my friend

Two photographs of the galaxy NGC 5253 show its normal appearance (left), with foreground stars in our Galaxy and the galaxy in 1972 when a supernova exploded (at top right). Such exploding stars can outshine their parent galaxies

Willy Fowler was still doing. But to have attempted to do so would have been a conceit, for over the past decade my path has not gone that way. I have never thought it necessary to have years of experience before entering the fray on a particular field, but I do think it necessary to be avid in the learning of up-to-the-moment details. It is better to be an on-duty novice than an off-duty expert, as John Edensor Littlewood might have said. Through the 1960s, to be on duty in the field of nuclear astrophysics it became more and more necessary to be closely associated with a major nuclear laboratory, the way that Willy Fowler has been at Caltech. Willy was always generous in arranging extended visits for me to Caltech, so for quite a number of years I was able to enjoy the same advantage. Administration in Cambridge and science policy at the Science Research Council over the years 1967–72 interrupted the arrangement, however, and when other pathways beckoned I took them. Hence the sadness one always feels at a might-have-been situation, like a child unborn.

Until the mid-1950s it had been generally thought the chemical elements were formed at a "big bang" origin of the Universe, but because of the variabilities of composition that were appearing in stars the stage was set in 1956 for a rejection of this point of view.

George Gamow tried a rearguard action, something connected with manganese, "mangan" as George always called it. He kept telephoning Caltech from La Jolla, eventually suggesting that Fowler, the Burbidges and I should go down there for a general discussion with himself and Hans Suess. We were puzzled, first by why George should be in La Jolla at all, and secondly by why he didn't simply come up to Caltech himself. We were intrigued to know the reason, especially as the explanations offered over the telephone seemed wildly confused, so we decided to make the trip. On that occasion we held our scientific meetings in one of the wooden huts, belonging to the Scripps Institute, which were then dotted picturesquely over a hillside overlooking the sea. Two or three of those huts were still there on my last visit to La Jolla but I fear that before many more years have passed by they will all be gone, swept away by the tide of progress, as one says.

So why had George Gamow been unwilling to drive 250 kilometres or so up to Caltech? Because he was spending two months in La Jolla as a consultant to General Dynamics. Not, it seemed, that the corporation asked him to do anything, but it did require him to be there each day in La Jolla, not necessarily at the corporation's buildings but somewhere in La Jolla, on the beach for instance. But could George Gamow not have forgone his consultancy fee for a day or two in order to visit Caltech, the persistent reader will ask? No he could not, for George had already blown every penny of his still-to-be-earned money, most of it in the purchase of an enormous white Cadillac convertible.

George's interest was more in cosmology itself than in the origin of the chemical elements, and there were times when George and I would go off for a discussion by ourselves. I recall George driving me around in the white Cadillac, explaining his conviction that the Universe must have a microwave background, and I recall my telling George that it was impossible for the Universe to have a microwave background with a temperature as high as he was claiming, because observations of the CH and CN radicals by Andrew McKeller had set an upper limit of 3 K for any such background. Whether it was the too-great comfort of the Cadillac, or because George wanted a temperature higher than 3 K, whereas I wanted a temperature of zero K, we missed the chance of spotting the discovery made nine years later by Arno Penzias and Bob Wilson. For my sins, I missed it again in exactly the same way in a discussion with Bob Dicke at the twentieth Varenna summer school on relativity in 1961. In respect of the microwave background, I was evidently not "discovery-

The horn antenna at Holmdel, New Jersey, with which Arno Penzias and Robert Wilson discovered the microwave background radiation in 1965. Most astronomers believe the radiation is a cooled-down relic from the big-bang origin of the Universe

prone" as Tommy Gold, an experimental astronomer who moved from the Royal Greenwich Observatory to become a professor of astronomy at Harvard University in 1957, puts it. Tommy tells the story of the manager of Henri Bequerel's laboratory who, on receiving an assistant's report (before the discovery of radioactivity) that photographic plates kept in the same cupboard as a certain bottle of salts were becoming fogged, simply instructed the assistant to keep the plates and the salts in different cupboards. This was just the way I felt about my previous conversations when the microwave background was actually detected in 1965.

It is widely believed that the existence of the microwave background killed the "steady state" cosmology, but what really killed the steady-state theory was psychology. Tommy Gold and Hermann Bondi had for long urged the predictability of their version of the theory as its outstanding virtue, and yet here, in the microwave background, was an important phenomenon which it had not predicted. Bad. For my part the situation was somewhat different. Holding a different view of the theory, I was more worried that it had acted to block the correct interpretation of the CH and CN observations. Those observations had been burned into my mind already in 1940, from the circumstance that the appearance of McKeller's work saved a paper by Ray Lyttleton and myself from being rejected by referees. The paper had predicted the existence of interstellar molecules, and fortunately for us the editor of the journal in question, A. H. Wilson, decided that a correct prediction should not be rejected. So it happened that I had had a big head start towards the microwave background and had then allowed a cosmological theory to cause me to behave like the manager of Becquerel's laboratory. For many years this knocked the stuffing out of me, again a matter of psychology.

It was already clear in the late 1960s that measuring the form of the microwave spectrum was not going to be an easy job, and so it has proved. The latest data differ by so much from what theory would suggest as to kill the big-bang cosmologies. But now, because the scientific world is emotionally attracted to the big-bang cosmologies, the data are ignored. What was sauce for the steady-state goose is not sauce for the big-bang gander. It is also a little bad for the big-bang cosmologies that nuclear astrophysicists are beginning to talk with some confidence of appreciable amounts of helium being synthesised in supermassive stars. If I were a supporter of the theory this drift of the evidence away from its two strong points would have me worried.

Energy appears in science in many guises, but through all the complexities one can distinguish a straightforward scale of increasing energy bindings, increasing depths of the energy well, as one might say. At the top of the well are the delicate interchanges of molecular structure that form the stock-in-trade of biology. Then come the bindings of atoms into individual molecules, the basis of chemistry. Then the bindings of electrons to the nuclei of individual atoms, atomic spectroscopy, the source of classical quantum mechanics. Atomic spectroscopy for the inner electrons of the heaviest atoms carries one to the energy-verge of nuclear physics, and so to the knowledge that has been crucial for the astronomical developments of the past quarter century. For a while there was the hope in the 1950s that the bindings of protons and neutrons yielded the lowest possible level of the energy well – hope because without hope it would have been impossible to summon the determination that physicists put into their search for an ultimately satisfactory nuclear model.

From the complications that piled up more and more in these studies, and from other heavy particles besides neutrons and protons present in cosmic rays, one could suspect that hope would not be requited; and with the early successes in the mid-1960s of the theory that the heavy particles consist of quarks it could indeed be seen that the energy well was likely to go very much deeper still. Today, it is apparent that the well may go as much as 10^{20} deeper, and that the so-called high energies of nuclear transformations are nothing but tiny adjustments within the much deeper well of quark physics. For many years, more than half a century, there have been hints to this effect from cosmology, from mysterious dimensionless numbers of order 10^{40} which appear remorselessly in all forms of theory. Just as from our present-day vantage-point we can see that the pioneers of stellar structure in the 1920s and 1930s were gravely handicapped by their lack of knowledge of nuclear physics, so it is likely that all cosmological studies to this point have been handicapped by a lack of knowledge of quark physics.

The young turks of the present day are very properly seeking to remedy this defect, with immense enthusiasm along what seems to me a wrong path. They are seeking to settle the hash of the whole Universe in the merest fleeting time interval following its origin. But the interesting quark transformations are almost immediately over and done with, to be followed by a little rather simple nuclear physics, to be followed by what? By a dull-as-ditchwater expansion which degrades itself adiabatically until it is incapable of doing anything at all. The notion that galaxies form, to be followed by an active astronomical history, is an illusion. Nothing forms; the thing is as dead as a door nail.

In retrospect, one can see that the existence of a high-energy world in astronomy might have been inferred already in 1932, from the discovery of cosmic radio noise by Karl Jansky. Yet astronomers were so keyed to the lower-energy near-thermodynamic world of stellar atmospheres that Jansky's discovery passed almost unnoticed. Nor did established astronomers expect much to emerge from radio astronomy in the years following the end of the 1939–45 war. Nor, I think, did anybody really expect the enormously prolific flowering of X-ray astronomy in recent years.

It is easy to pass off these misjudgements by the precept that everything in nature turns out to be much richer than one expects or by the cruder statement that in the process of becoming "established" scientists tend to develop into fuddy-duddies. While there is an element of truth in these generalities, I think it more correct to say that gross and persistent misjudgements imply the overlooking of some crucial point of principle, as for instance Lord Kelvin overlooked the nuclear level of the energy well in his famous controversy with geologists over the age of the Earth. Like Lord Kelvin, I suspect astronomers misjudged the potentialities of radio astronomy, X-ray astronomy and the like, because they overlooked the quark level of the energy well. Unaware of the existence of the quark level, they passed off each new discovery in little better than a hand-waving style, convincing themselves that each case was some kind of a freak, and that no more of them would be found.

Not everyone was comfortable with the situation, however, and nobody fought harder against it in my experience than Geoffrey Burbidge. It is now just a quarter of a century since Geoffrey convinced me that the nuclear level was inadequate to explain the wide range of emerging high-energy phenomena. But what deeper level could there be? All we could think of in 1956 was matter–antimatter annihilation. Nothing that was much good came of this idea, and I am glad now that instead of hand-waving about it we admitted straight out that it didn't work.

The next possibility was gravitation, and gravitation it has been which has captured the attention of most astronomers over the past 15 years. Is this not adequate the reader may ask? Why bother with the quark level of the energy well? Because gravitation is almost surely a phenomenon that itself belongs to the quark level – there is nothing at either the electromagnetic or nuclear levels which explains gravitation (think how hard Einstein tried at the electromagnetic level). I suspect gravitation could look very different in a generally stirred-up quark picture, which is why I distrust the application of simple gravitational theory, the empirical theory, to high-energy phenomena. Although prediction is at best a risky profession, I will offer the prediction that the understanding of high-energy astronomy and the understanding of gravitation will turn out to be in one-to-one relationship; you cannot have the first without the second, and moreover the second will guarantee the first.

Let me explain a point that is often overlooked about violently explosive phenomena. It is a common mistake, repeated many times in the literature, to think that because an explosion produces a lot of energy the energy is necessarily observable. Most of the energy of an explosion usually goes into a system of particles that eventually move uniformly outward from the centre of explosion, and the energy of uniform motion cannot be observed. It is only accelerations and decelerations of particles that are observable; what one might call the internal motions of the particles. Going along with the stream does nothing, just as sticking to conformist opinions does nothing for one in human society. This is not my punch line. The punch line is that, even though outward speeds are maintained in a free explosion, internal motions are not. Internal motions die away adiabatically, and the expanding system becomes inert, which is exactly why the big-bang cosmologies lead to a universe that is dead and done with almost from its beginning.

The intense richness of the high-energy aspects of the Universe as they have been revealed over the past 25 years must therefore be due to a developing picture, and it is this that the new generation of theorists should be attacking, not a by now hopelessly expanded residue from a hypothetical beginning of the Universe.

In retrospect, I do not feel that those of us who argued for the steady-state cosmology put our case in the most appropriate light. I tried too hard to give a mathematical and physical description of the theory, at a time when the essential physics was unknown. Herman Bondi and Tommy Gold tried for a broad principle, but even so made a choice that turned out to be too restrictive. The position I take today is that the wider principle of uniformity of the geologist James Hutton is as true in astronomy as it is in geology – everything we observe is related to continuing processes. Just as it was unnecessary for Hutton to be aware of plate tectonics, so the principle applied to astronomy does not require one to guess still-unknown physics. If we had taken this position back in the 1950s, a position much in the spirit of what we were then trying to say, I think we might have had a better audience, and certainly we would have had a better chance of riding out the subsequent storms.

It would be possible to compile a long list of famous observational discoveries made over the past quarter century, each an interesting story in itself. If one were restricted to only three, which would one choose? Restricted to three more, as I have mentioned the microwave background already, from so remarkable a list there is ample room for one judge to differ from another. My own choices would be the quasars, which had been "in the air" since 1961 but which were found explicitly in 1963 due to a combination of the work of Cyril Hazard and Maarten Schmidt; pulsars discovered in 1967 by Jocelyn Bell and Tony Hewish; together with the cooperative discovery by many astronomers of organic molecules ubiquitous in interstellar space.

I recognise here an unrestrained bias toward ground-based astronomy. For me, astronomy conducted by satellites in space is painted on too large a canvas for individual accomplishments to have quite the intense impact of the particular discoveries I have just mentioned. For others it is almost surely not so, and for the future it will certainly not be so. A quarter of a century hence, the situation may well be reversed, with most, if not all, the outstanding achievements belonging to space astronomy.

There was a time in the 1960s when NASA was anxious to secure the cooperation of "established" astronomers, and I have sometimes wondered how things would have stood today if I had responded positively. While it is true that I had more than enough other things to do, this was not my main reason for a lack of interest in space research at that time. I could not accommodate myself to scientific programmes playing second fiddle to the astronaut programme, which I regarded as a hugely expensive luxury. There were plenty of scientists in the United States who thought likewise.

Some astronomers took the same line as I did, while others argued that as the opportunity existed it should be seized, irrespective of distortions of priority. Whenever I ponder the marvellous results achieved in recent years from various unmanned planetary missions I always turned out to be right. Does it pay therefore to sup with the devil, I keep asking myself? Perhaps it does where science policy is concerned. Perhaps in astronomy one should adopt the Faustian policy of selling one's soul for a telescope. Or "my kingdom for a telescope" as an enlightened Richard III might have said. For unless the skies are open, astronomy is an arid science. In this respect the position of young British astronomers today is incomparably better than it was at the beginning of the quarter century, thanks to the major telescopes that are available to them.

What of the future? Experience shows that it is never correct to answer this by saying "more of the same". The Universe always turns out to be incomparably more subtle than we expect, which is exactly why I continue to be obstinately doubtful about the dead-and-done-for universe of the big-bang cosmologies. Consider how almost every aspect of every planetary mission throws up an unanticipated surprise. Every little satellite has more "go" left in it than the exhausted models of those cosmologies.

Dare I suggest cosmic biology? I don't know how long it is going to be before astronomers generally recognise that the combinatorial arrangement of not even one among the many thousands of biopolymers on which life depends could have been arrived at by natural processes here on the Earth. Astronomers will have a little difficulty at understanding this because they will be assured by biologists that it is not so, the biologists having been assured in their turn by others that it is not so. The "others" are a group of persons who believe, quite openly, in mathematical miracles. They advocate the belief that tucked away in nature, outside of normal physics, there is a law which performs miracles (provided the miracles are in the aid of biology). This curious situation sits oddly on a profession that for long has been dedicated to coming up with logical explanations of biblical miracles.

Actually, biblical miracles have a certain interesting quality about them. I suspect, for instance, that the miracle of the loaves and fishes wasn't written by a rogue or a nit-wit but by a person who had seen, or who knew someone who had seen, a beached whale. And if such an incident happened in the childhood of the author, the later writing of the story would not have seemed to make an unwarrantably serious concession to logic. It is quite otherwise, however, with the modern mathematical miracle workers, who are always to be found living in the twilight fringes of thermodynamics.

I would have despaired of being able to bring this point home in its full enormity to even the modern expensively educated public had it not been for Professor Ernö Rubik. I wonder how many parents, after watching their children fiddling with a Rubik cube, have decided to have a go themselves, only to receive the shock of their ageing lives. Of course it is not hard to solve a scrambled Rubik cube if you follow written instructions with the meticulous precision of an accountant, but to solve a scrambled cube given only the cube – no instructions, written or verbal – is something of a corker. At least it took me three days of not inconsiderable effort, plus three more days before I could be certain of solving any of the 4×10^{19} possible scramblings of the cube. Although I have since managed to get the time for a solution down to about 10 minutes – still pretty poor to an expert – I haven't managed to figure out the group theoretic structure to which the cube belongs. Unlike the rotation group with only three independent rotations, the darned cube has six rotations, which knocks your mathematician straight out of the ground from the word go.

At all events, anyone with even a nodding acquaintance with the Rubik cube will concede the near impossibility of a solution being obtained by a blind person moving the cube faces at random. Now imagine 10^{50} blind persons each with a scrambled Rubik cube, and try to conceive of the chance of them all *simultaneously* arriving at the solved form. You then have the chance of arriving by random shuffling of just one of the many biopolymers on which life depends. The notion that not only the biopolymers but the operating programme of a living cell could be arrived at by chance in a primordial organic soup here on the Earth is evidently nonsense of a high order. Life must plainly be a cosmic phenomenon.

Quite a number of my astronomical friends are considerable mathematicians, and once they become interested enough to calculate for themselves, instead of relying on hearsay argument, they can quickly see this point. Then they become happy to take over normal biology, Darwinian theory and all, into a cosmic setting; and if this is the way it eventually turns out the consequences for astronomy will be far-reaching. Yet my thoughts run on to more fantastic possibilities, at which my friends usually blanch and turn away. The dividing line here is interesting. If one simply transfers present theory to a cosmic setting, then astronomy is destined to swallow up biology, but if the more fantastic possibility is correct the reverse will be true, which convinces me that the more extreme possibility will find eventual favour with biologists.

PART TWO

The Roots of Astronomy Flourish Again

The science of astronomy stretches back to at least 5000 years ago, and these deep roots provide a rich nourishment for the flourishing of astronomy today. The heritage of the early star-gazers of the Mediterranean permeates our lives to such an extent that it is hard to see in true perspective. Our names for the planets and the patterns that we recognise as constellations have descended – through the classical civilisations – from the ancient peoples of Mesopotamia, at least 4000 years ago. Even more basic, our familiar calendar and 24-hour division of the day are legacies from the astronomical observations of the earliest scientific civilisation, that of the Egyptians of the Nile valley, who were using this system before 3000 BC.

The cultural heritage from civilisations all around the world also contains a surprising wealth of astronomical data useful to modern science. In recent years, astronomers have devoted considerable effort to sifting out these gems of observation from a diversity of sources, from Japanese court diaries to broken clay tablets in the ruins of Babylon. An observation may relate the appearance of a "guest star", now known to be a supernova whose remains can be located from these ancient records. It may describe the blemishing of the Sun by a large sunspot, an observation that gives clues to the past behaviour of the Sun's cycle of sunspot activity. Or an entry in an ordinary diary may be prompted by the wonder of a total eclipse of the Sun, and our resulting knowledge of the date and location of the eclipse can lead to important deductions about the Earth's rotation, and hence about our planet's interior.

We cannot, however, merely delve back through the written records of any civilisation as we please to find data on all these celestial phenomena. It is a remarkable fact that all the most important civilisations have seen the sky in different ways, concentrating their attention on some kinds of object or phenomena, and ignoring

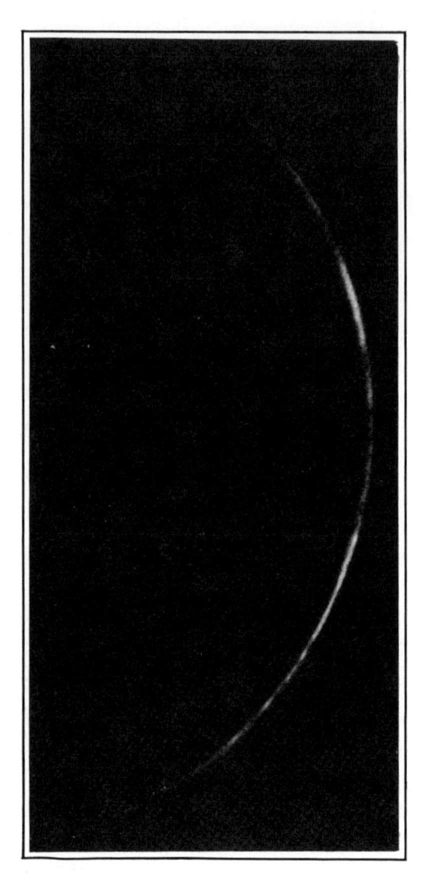

The very first appearance of the crescent Moon was used by many ancient civilisations for regulating their monthly calendar – one of the earliest practical uses for astronomy. Here the Moon is seen only 21¼ *hours "old", as one of the narrowest crescents ever photographed*

others that now seem equally obvious and important. The only phenomena consistently recorded everywhere are eclipses.

Take for example two of the most important sources of astronomical data, the Babylonians and the Chinese. Both civilisations observed the sky primarily for astrological purposes. The Babylonians were concerned almost exclusively with the motions of the Moon and the planets, and they used their considerable mathematical ability to predict the paths of these "wanderers". From this has descended not only the science of planetary orbits, but also today's "pop astrology" with its predictable planets moving through the Zodiac. The Babylonians had correspondingly little interest in unpredictable phenomena – the appearance of meteor showers, comets, novae and supernovae. Chinese astronomers, however, were content merely to observe the positions of the planets for their immediate astrological value, making no attempt to predict their future positions. Indeed, their astrology was based on the unexpected. In a Chinese observatory several astronomers would watch the sky all night to ensure that no sudden portent would escape their attention: they recorded aurorae, "thunderlike noises", the positions of the Moon and planets when they were near each other or important star groups, comets, meteor showers and the appearance of "guest stars" – novae and supernovae. This is clearly a rich vein for modern astronomers to prospect.

Even near-neighbours could differ in their main astronomical interests. In America, the Aztecs of Mexico and the Incas of Peru were followers of the Sun, erecting observatories lined up with the Sun's rising and setting, or equipped with vertical tubes down which the Sun would shine when overhead. But the central American Maya based their astrology on the planet Venus, with at least one observatory aligned to the extreme northern and southern points at which the planet sets.

Some civilisations did not regard the heavens primarily as an astrological display. To the Polynesians, the brightest stars were in the sky to aid navigation from island to island. The Egyptians were mainly concerned with using stars – particularly Sirius – for time-keeping and for regulating the calendar, even though the skies were also the home of the deities. Time-keeping by the stars has remained part of the astronomer's job to this day, as witnessed by the hourly "six pips" from the Royal Greenwich Observatory.

These diverse sources of information have each, to a greater or lesser extent, enriched modern astronomy, with the greatest scientific return so far coming from the observations of comets, sunspots, supernovae and eclipses. But much remains to be done, in the analysis of known records and the discovery of new sources. Even the great stone circles and rows of western Europe – Stonehenge on Salisbury Plain and Carnac in Brittany – may be silent repositories of knowledge about the Moon's orbit and the Earth's changing tilt, if, as some astronomers believe, they were accurately aligned on the rising and setting of the Sun and Moon. Less speculatively, much more written material must exist. Unfortunately, the Spanish conquerors destroyed almost all the native American records, but we have still retrieved only a fraction – perhaps one-twentieth – of the clay tablets from Babylon. Even after its successes so far, historical research in astronomy undoubtedly has a great future.

2

How the stars kept time in ancient Egypt

ED KRUPP

3 January 1980

From the 4th century BC Egyptian astronomers looked for order and regularity in the heavens. They used the observations in everyday life, and from them we have inherited the 12-month year and the 24-hour day.

Along 1300 kilometres of Egyptian Nile may be found the tatters of a civilisation so taken with the sky that the imprint of astronomy remains indelible even after 5000 years. To the Egyptians the sky was a rich metaphor. Its patterns of life, growth, death and rebirth were stencilled onto the Earth. The cycles of human life resonated with the sky.

Egyptian astronomy was inextricably entwined with Egyptian religion. The Egyptians plucked images from the sky and incorporated them into their temples, tombs and tales. New Kingdom temples, at Karnak and Abu Simbel, seem to be aligned with the Sun. Paintings in the royal tombs of Thebes have numerous astronomical connotations. Even Giza's Great Pyramid is linked with the sky.

A more practical product of the Egyptian astronomers was a solar civil calendar, based upon the length of the tropical year. This scheme, and the Egyptian 24-hour system of time-keeping, evolved into the calendar and clock used today. Curiously, these two vital components of organised society are most apparent in the paraphernalia of the priests and the dead. What we know of Egyptian astronomy is, in large measure, limited to what is found in temples and burial grounds.

We see in Egypt the *applications* of astronomy. The astronomy itself is only implied, and the monuments suggest Egyptian astronomy was a lightweight science. There is no evidence of a technical vocabulary. There is no written record of systematic observations. The comparison is usually drawn with Mesopotamia, where astronomy was mathematical and bordered on true science. This is misleading, however, for mathematical astronomy in ancient Babylonia may have begun as late as the 7th century BC. By contrast, when we first think of Egypt, anything from the Old Kingdom pyramids of the third millennium BC to the Ptolemaic temples, about 2500 years more recent, may come to mind. If we are really intent on evaluating relative expertise, we must be clear about what Egypt and what Mesopotamia we have in mind.

Whether astronomy in Egypt was mathematical or scientific in a sense hardly matters. Astronomy percolated through Egyptian culture, and to appreciate its flavour we have to see how the Egyptians interacted with the sky.

Egyptian astronomers were time-keepers. It was their function to monitor the sky to ascertain the hours and keep the calendars. One very late text, the *Papyrus Carlsberg 9*, exists that seems to have been written by an anonymous astronomer. It provides a table and technique for calculating the date, in the Egyptian civil calendar, of the start of each lunar month, in a 25-year cycle. The document is no earlier than AD 144, but Richard A. Parker, a specialist on Egyptian calendrics, has shown that the text reflects a tradition 500 years older than the papyrus itself.

Hieroglyphic inscriptions are our most direct sources of Egyptian astronomy, but little else is as explicit as *Papyrus Carlsberg 9*. Before Jean François Champollion cracked the hieroglyphic code in the early 19th century, classical commentaries, often unreliable, defined Egyptian astronomy. Once hieroglyphics could be read, at least in part, inscriptions were copied, collected and translated by a variety of scholars and travellers. In an ambitious and successful expedition (1842–45) Karl Richard Lepsius, a German, acquired enough material from Egyptian monuments to attempt, in 1849, a chronology of Egyptian history. Lepsius also recognised and translated the astronomical character of many texts.

Some astronomical texts were, in fact, applied later to the problem of Egyptian chronology. By associating an event mentioned in an ancient inscription with a calculated date, the ancient chronology – what happened when – was tied down. References to heliacal, or the first pre-dawn, risings of Sirius, a star of particular importance to the Egyptians, were linked by their contexts to specific dates.

Heinrich Brugsch, also a German and a renowned Egyptologist, translated many inscriptions and collected them into a *Thesaurus Inscriptionus Aegyptiacarum* in five volumes. The first number in the series, *Astronomical and Astrological Inscriptions on Ancient Egyptian Monuments,* was published in Leipzig in 1883 and remained for many years the fundamental work on Egyptian astronomical texts. More recently, other researchers, among them Alexander Pogo, Otto Neugebauer and Richard A. Parker, have analysed these and other texts in much greater detail, and comprehension of Egyptian astronomy has improved accordingly.

It was obvious, by Brugsch's time, that the Egyptian calendar and, in particular, the New Year were calibrated by the reappearance of Sirius in the sky before dawn. Sirus, like most stars, spends part of the year – 70 days, in fact – in the daytime sky. When the Earth moves in its orbit about the Sun, the Sun appears to move eastward relative to the background stars, which are, of course, invisible by day. Eventually the Earth will travel far enough to permit Sirius, or any other temporarily invisible star, to rise just long enough before the Sun to appear, at least briefly, in the pre-dawn sky.

The reappearance before dawn, or heliacal rising, of Sirius was obvious to Egyptian astronomers. Sirius is the brightest fixed star in the sky, and its heliacal rising occurred close to the summer solstice and preceded the all-important Nile Flood. The Egyptians counted three seasons and based them upon the behaviour of the Nile. The seasons – "flood", "emergence" and "low water" – were each composed of four lunar months. Ancient texts provide the names of these months, and reliably timed feasts occurred in each of them.

Although the Egyptian year was bound to the Nile, its months, if unattended, could quickly outpace the seasons of the tropical year. A "year" of 12 lunar months is 354 days, approximately 11 days short of the solar year. The early Egyptians solved this problem, however, by intercalating a month whenever the heliacal rising of Sirius occurred in the last third (or 11 days) of the last lunar month. The name of the intercalary month was Thoth, and there is evidence for use of this technique at least as early as the time of the building of the Pyramids.

Early in the third millennium BC another calendar was introduced. Administration of government and commerce presumably stimulated a more regularised system based on a solar tropical year of 365 days. The old calendar based on the Moon and stars continued to determine the liturgical activities, at least for a while. Civil matters became the jurisdiction of the new solar calendar. The true length of the tropical year is closer, however, to 365¼ days. With such a discrepancy, the civil year's original synchronisation with the lunistellar calendar dissolved, and a new lunar calendar was inaugurated. Its intercalations were timed by the civil new year. Even so, the civil year slowly drifted through the seasons, with a period known as the Sothic cycle, perhaps 1461 Egyptian civil years long. Rather than abandon any of the three calendars, the Egyptians maintained them all, even into classical times.

The civil year was a sensible innovation. Its "months" were not months at all, for they were not really determined by the phase of the Moon. Instead they were set to be simply 30 days each. Twelve 30-day months plus an additional five "epagomenal" days at the end generated a tropical year whose passage could be timed with the heliacal rising of Sirius. So there had to be Egyptian astronomers – someone had to watch that star.

Each of the "months" in the civil year was subdivided into three 10-day intervals called "decades". Just as the pre-dawn reappearance of Sirius signalled the new year and the start of the first decade of the first month, a sequence of stars, known as "decans", were chosen to indicate the commencement of each of the other decades of the year. These decans were used also to indicate, by their risings, the various hours of the night.

Julius Caesar assigned an Egyptian, Sosigenes, to convert Rome to the solar civil calendar, and through Rome this calendar became universal. We take the calendar for granted today, and yet it is a small miracle that two people can make an appointment and meet in the same place at the same time on the appointed day. Contrary to an often repeated claim, a systematic calendar is not so essential for successful planting and harvesting. This is, by now, a rather tired explanation for the early development of astronomy. The distribution of goods and services and the interaction of society are different matters altogether, however. The calendar's real power is the organisation and stability it imposes on human activity. The calendar is a fundamental component of culture, and we have the Egyptians to thank for ours.

Egyptian astronomers left as a second legacy the 24-hour day. Time-keeping, like the calendar, is a basic tool of organised society. The Egyptians marked the passage of the night by the risings of stars. These were the same 36 decans whose heliacal risings announced the year's decades.

The special significance of Sirius prompted the Egyptians to choose as decans stars which mimicked the period of invisibility and pattern of reappearance of Sirius. Of course, at any given time of year only a certain number of decans would rise during the course of a night, but each, as it rose, indicated the start of another "hour". It might be thought, with 36 decans around the sky, one-half of them, or 18, might be observed to rise during the night. This in turn would suggest an 18-hour night. Actually, however, the length of complete darkness varies seasonally. The Egyptians calibrated their calendar at about the time of summer solstice, when nights are shortest and only 12 decans appear.

At first, the Egyptian "hours" varied in duration from one time of year to another. Later, however, hours of equal duration were instituted, and a symmetrical set of 12 daytime hours (including morning and evening twilight) was introduced. The entire scheme is a direct consequence of the solar civil year, its decimal (10-day) subdivisions, and its calibration by the heliacal rising of Sirius in midsummer at the Nile's flood. The Egyptians made these choices, all very sound and consistent and, to them, meaningful, and as a result people now keep to a 24-hour day.

The earliest evidence of a 12-hour night dates from Egypt's Middle Kingdom. On the undersides of some coffins belonging to the time from the 9th and the 12th Dynasties (2160–1786 BC) appear an arrangement of decan names in a grid-like array. These designs have come to be known as diagonal star clocks after the reappearance of decan names in adjacent columns, along diagonal lines.

The diagonal star clocks are not clocks at all but simply diagrams, or tables, indicating which star marks each hour of the night during each of the 10-day intervals. The Earth's gradual movement around the Sun shifts each decan's position, and therefore time of rising, with respect to the Sun. The star that the *imy-wnwt*, or "hour-watcher", saw at the 12th and last hour of the night will rise at the 11th hour 10 days later. Each decade the star slips back an "hour", and its name leaves a diagonal trace on the coffin lid.

By the era of the Ramesside pharaohs (1300–1100 BC) the watching of decans had shifted from risings to transits. The evidence still appears in a funereal context – in this case, a diagram of star names and positions shown in association with a seated figure. These star clocks appear in the royal tombs of the Ramessides.

Missing, if indeed they existed at all, are the actual papyri used by the Egyptian astronomers to keep track of which stars did what when. But the decanal system was important. Inclusion of these star clocks in the environment of the dead suggests that knowing the hour and the date was as vital for the transit to the next life as for the passage through this. Prayers or charms – a right word at the appropriate time – protected the dead.

The sky's drama was equated with human destiny. The flavour of this is apparent in the interchangeability of astronomical language and funereal language. When a decan star disappears for 70 days, it is said to die and enter the embalming house of the underworld, Duat. It remains there for 70 days and loses its impurities. Then, it is reborn in the east. Similarly, a body was embalmed for 70 days prior to burial. The underworld itself was divided into 12 regions, one for each hour of the night. It was the Sun's destiny to enter Duat at sunset, pass through the perilous realm, and emerge victorious at dawn. Accounts of this symbolic journey appear as religious texts in the royal tombs of Thebes. There also are portrayals of the sky goddess, Nut. The Sun disc at her mouth symbolises sunset. She swallows the disc, and it passes through her body. For the dawn, it reappears at her loins, literally born from the sky. The daytime disappearance and nightly rebirth of the stars is shown in the same way.

One of the most persistent astronomical images of ancient Egypt is the set of constellations known as the Northern Group. This unlikely collection of celestial residents includes, among others, a bull (or often just its foreleg), a hawk-headed human, sometimes shown with a spear, and an upright female hippopotamus, sometimes cloaked with a crocodile. In some portrayals the hippopotamus leans upon an enigmatic device. It looks like a knife, is called a mooring post, and represents the North Pole of the sky. The stars of the Northern Group are the circumpolar stars. To the Egyptians they were the "imperishable" or "undying" stars, for they never passed below the horizon into Duat.

The group of stars that form the Big Dipper, or Plough, was the bull (or its foreleg). Its circuit about the pole was symbolised by the tethers which continued down to the mooring post. The hippopotamus was composed, at least in part, of some of the stars of the constellation Draco, the Dragon. Thuban, the star nearest the pole in the Old Kingdom, may have been associated with the mooring post.

Although much has been made of the Great Pyramid's supposed symbolic significance, the "imperishable stars" help confirm that it was, indeed, the tomb of Khufu, the 4th Dynasty pharaoh who built it. The sides of the Great Pyramid are aligned with great accuracy to the cardinal directions: north, south, east and west. These, after all, are astronomical, for they derive their meaning through the

The northern group of constellations, as painted on the tomb of Seti I, include a female hippopotamus (in western tradition part of the constellation Draco) holding a mooring post which represents the North Pole of the sky. The bull tethered to the mooring post is our Plough (Big Dipper)

apparent daily rotation of the sky. The worst error in the alignment is only 5½ arcminutes, just about one-sixth the size of the full Moon as we see it in the sky. This remarkable accuracy is maintained in an artificial mountain of 2.3 million stone blocks, 230 metres long at the base on a side and originally 146 metres high.

Inside the Great Pyramid is a mystifying collection of chambers and passages. One of these, the King's Chamber, still contains the empty, lidless remains of a granite sarcophagus. Two curious openings may be found just a metre or so above the floor and toward the east end of the room, one on the north wall and one on the south. They are the ends of two shafts. Both extend horizontally and then bend upward. The north shaft is set at an angle of 31 degrees, the south at 44 degrees 5 minutes. The shafts continue at these angles nearly to the outer faces of their respective sides before they bend horizontally again and open to the sky.

The usual explanation for the King's Chamber "air shafts" has been ventilation, but the Egyptians were better at air-conditioning than this. In any case, this is a very difficult way to achieve the presumed effect. Egyptologist Alexander Badawy, of the University of California at Los Angeles, assisted by an astronomer, Virginia Trimble, of the University of California at Irvine, demonstrated in 1964 that the north shaft pointed toward the upper culmination of Thuban's circumpolar arc in 2650 BC, when the Great Pyramid was built. Similarly, the south shaft was oriented on the transit of Orion's belt, perhaps on its central star, Alnilam.

These astronomical alignments need not be ascribed to chance, for Badawy cited written evidence to explain them. Inside the pyramid of Unas (5th Dynasty, 2350 BC) at Saqqara, the walls are completely carved in hieroglyphics. These writings are the *Pyramid Texts*, and they describe, among other things, the pharaoh's destiny. He is to ascend to the imperishable stars and manage their appointed rounds. He is to ascend to the sky to join Osiris, the god of resurrection, as well. The Great Pyramid's northern shaft was not a sighting tube for Thuban. It symbolised the ascent of the pharaoh to the realm of the undying stars. In the same way the southern shaft represented the pharaoh's journey to Osiris. Osiris, in the Egyptian sky, was the constellation Orion. Astronomy, funeral rites and religion were combined in the Great Pyramid on monumental scale.

Astronomical allegory is a fundamental constituent of Egyptian religion. Some temples are covered with astronomical inscriptions and imagery. At Dendera's Temple of Hathor the new year festival was celebrated on the roof in an open kiosk. The temple is Ptolemaic and relatively recent (125–30 BC), but some of the traditions its walls preserve are much older.

Sir Norman Lockyer, the renowned astrophysicist and founder editor of *Nature*, was the first to argue that Egyptian temples had been precisely aligned on the risings and settings of the Sun and bright stars. His measurements and conclusions were published in 1894 in *The Dawn of Astronomy*. The quality of the work was mixed.

In 1973, Gerald Hawkins, a British-born astronomer known, like Lockyer, for his astronomical interpretation of Stonehenge, approached the problem of New Kingdom temple alignment more modestly. His results are more convincing. Karnak, near modern Luxor and 720 kilometres south of Cairo, is the site of the Great

Temple of Amon. The temple is huge and is but a part of a gigantic complex of temples. Construction and modification of the Great Temple of Amon continued over several centuries, but most of the important building belongs to the 18th Dynasty (1558–1303 BC). Its northwest/southeast orientation led Lockyer to believe that the temple was targeted on sunset at summer solstice. Hawkins by contrast has shown that an alignment with sunrise at winter solstice is more likely. He associates this astronomical event with a solar sanctuary possessing a window that opened to the mid-winter sunrise in the southeast. The sanctuary's axis is aligned with that event, and Hawkins cited some metaphorical texts to support the idea.

The principle of Karnak's "High Room of the Sun" is duplicated on a smaller scale 500 kilometres farther south in the rock-hewn temple of Rameses II at Abu Simbel. There the main temple appears to have been directed to the sunrise on the civil new year on the occasion of the 30-year jubilee of Rameses' reign. This date coincides with 18 October in the modern calendar. A small, easily ignored little chapel is situated at the north end of the main temple's monumental façade. Architecturally this chapel resembles the High Room of the Sun, and it is skewed to the winter solstice sunrise.

There is no doubt about the existence of ancient Egyptian astronomers, and astronomy's applications deeply penetrated Egyptian life. We relive their achievements in present-day calendars and clocks, although there is little direct evidence of the astronomers themselves. Amenemhet, of the 16th century BC, is the earliest of the few known by name. It is the tomb of Nakht, an astronomer of the 18th Dynasty, however, that offers the most convincing proof of the presence of an astronomer's lifestyle. The royal Theban tombs in the Valley of the Kings are rich in astronomical associations and imagery: the Northern Group, star clocks, representations of Sirius, Orion and the planets, and more. Nakht, by contrast, decorated his tomb with those things closest to an astronomer's heart – enthusiastic winemaking, sumptuous banquets and women as dancers and musicians, delicate and scantily clothed.

3

The astronomical riches of Chinese astrology

RICHARD STEPHENSON

26 June 1980

The Chinese bureaucratic system has left almost complete records of the sky's appearance, from 2000 years ago to the early part of the 20th century. Although made for astrological purposes, these observations provide modern astronomers with an invaluable sequence of sky phenomena – once they have translated into astronomical terms the appearance of "guest stars", "broom stars" and "flowing stars"!

Star-gazers in China, right up to the foundation of the modern republic in 1912, kept their night-time vigils to make detailed astrological predictions, not to improve their understanding of the Universe. During the most recent dynasty – the Ch'ing – the Chinese were making practically the same type of observations as their predecessors did in the first major dynasty – the Han – which began in 206 BC. Although this stagnant outlook meant that Far Eastern astronomers made little progress in understanding the Universe, they collected over the centuries an enormous variety of astronomical data, both in China and in its cultural satellites Korea and Japan. The observations of the astronomers – who were civil servants – were so important that they were included along with affairs of court and state when the official history of each dynasty was written up.

Undoubtedly the Chinese astronomers were first and foremost *observers*. However, a number of fundamental discoveries due originally to the Greeks, were made independently by the Chinese, although somewhat later. Examples of this are the realisation that the Moon receives its light from the Sun (2nd century BC or earlier in China) and the correct explanation of eclipses (1st century BC or earlier). Further, among the rival cosmological theories of China, some of which were very primitive, there is a minority school of thought with a remarkably modernistic outlook. Proponents of the *Hsuan-ye* ("Infinite Empty Space") cosmology found no difficulty in accepting the idea that stars and planets float freely in an infinite void; the notion of solid crystalline spheres to carry the various celestial bodies was purely a western concept.

The Chinese failed badly by European standards in their lack of any real attempt to understand the motions of the Sun, Moon and planets. There is no evidence of a constructive theory, either geocentric or heliocentric. This was probably mainly due to the absence of deductive geometry – in striking contrast to the Greeks. The astronomer Yu Hsi in the 4th century AD admirably crystallised the attitude of the Chinese towards planetary theory. He remarked: "The heavenly bodies are scattered about, each of them having its own course like the tides and waves of the sea and rivers and the movements of the numerous living creatures." In this view the motion of any particular luminary was its own peculiar characteristic, and there was no point in seeking any further explanation. There was no Claudius Ptolemaeus (Ptolemy) or Nicolaus Copernicus in the Far East. Because of the absence of any viable planetary theory in China, astronomers limited their predictions to the crude use of numerical cycles derived from earlier observations in forecasting eclipses. In marked contrast the Babylonians, who likewise had no concept of the planetary motions, developed the use of astronomical cycles to such an extent that they were able to predict, with tolerable accuracy, the motions of the Sun, Moon and planets into the future.

The role of observational astronomy in China, and its importance now, is intricately interwoven with the country's history. Chinese history, as distinct from myth and legend, begins surprisingly late compared with that of some of the great western civilisations. The historical period starts somewhere around 1500 BC, but for a variety of reasons there is precious little in the way of history before about 700 BC and certainly not as much as we might expect until as late as 200 BC. Figure 1 gives a list of the various Chinese dynasties together with details of some of the main events in observational astronomy.

Much of the ancient historical data of China was systematically and ruthlessly destroyed by command of the first emperor, Ch'in Shih-huang, in 213 BC. This notorious "Burning of the Books" was designed to sever all ties with the past and hence destroy the power of the feudal princes. In this it was highly effective, but as a result irreparable gaps have been left in the history of China before this time. Lately Ch'in Shih-huang has come into the news following spectacular excavations at an annexe of his tomb near Hsi-an. History records that the main vault was on a grand scale and filled with treasures, some of astronomical interest. Szu-ma Ch'ien, perhaps the greatest Chinese historian, writing in the 1st century BC, describes the scene within as follows. "The 'hundred rivers' of the Empire, the Kiang, the Ho and the great sea were modelled in quicksilver. Some machines made it flow and conveyed it to and fro. Above everything was the starry vault; below was the map of the Earth." There is some evidence that robbers penetrated the sepulchre

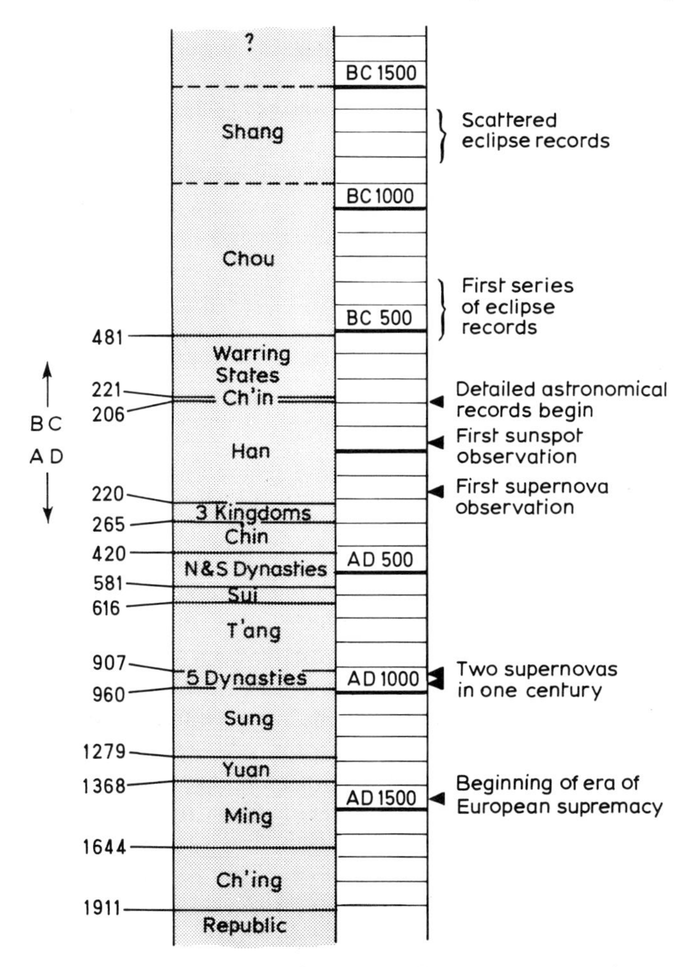

Figure 1 *The succession of Chinese dynasties covered over 3000 years. Some of the important astronomical events recorded in China are shown on the right*

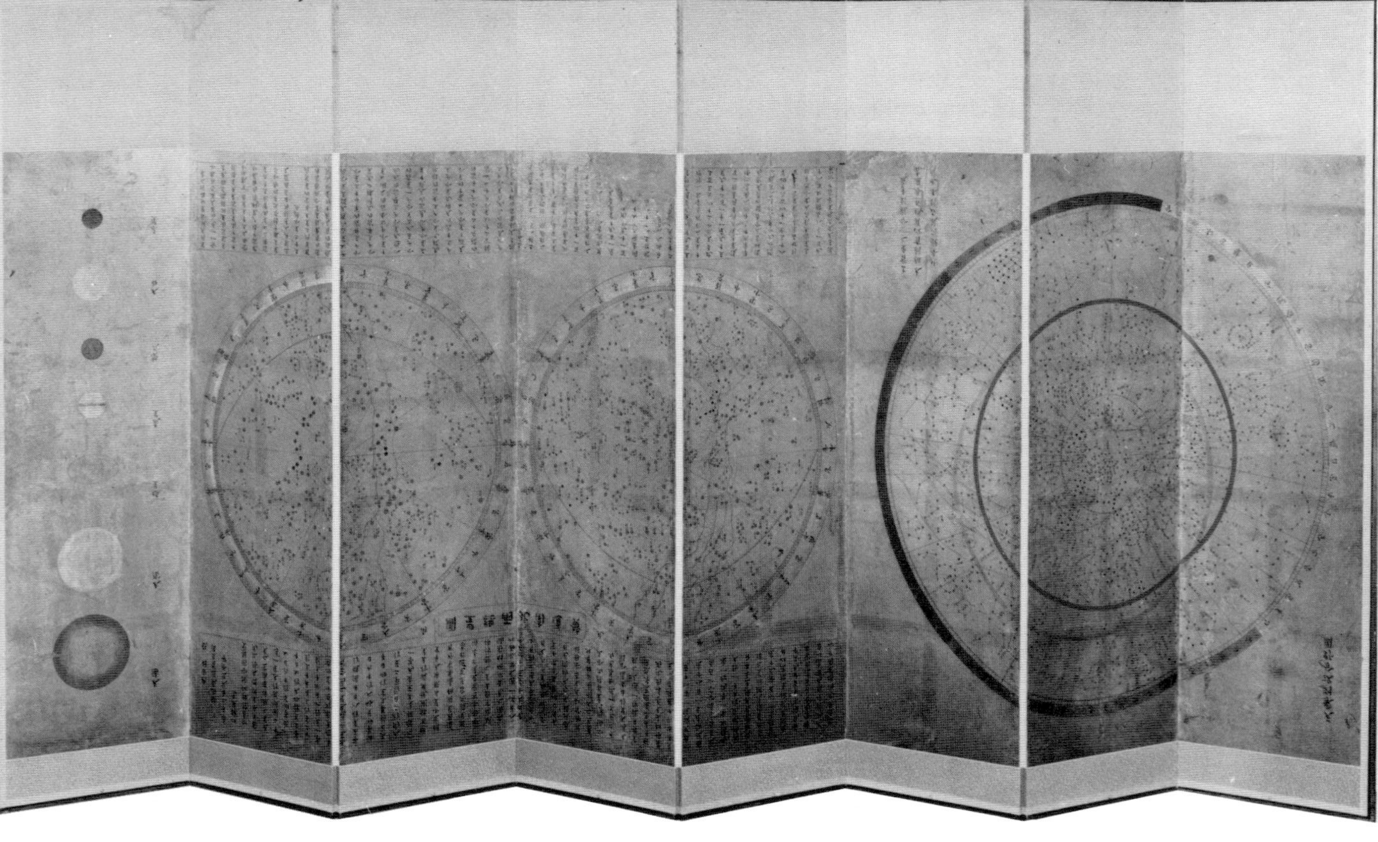

An 18th-century Korean screen depicts the traditional Chinese constellation patterns. The long connected lines of stars are a characteristic form for Chinese constellations

in antiquity. However, just what literary and other treasures it may hold awaits further excavations.

Referring to Figure 1, the first truly historical – as opposed to mythical – dynasty was the Shang (1500–1000 BC). At the beginning of the present century, peasant farmers accidentally discovered vast numbers of inscriptions from this period at the site of one of the Shang capitals (near the modern city of An-y ang). These were written in a very primitive form of Chinese script on materials such as tortoise shells and ox shoulderblades. Unfortunately, interesting as they most certainly are, they tell little of the history of the times for they are essentially oracle texts. However, they clearly indicate some form of organised astronomical activity at this very early period. There are occasional references to eclipses and certain stars on the oracle bones, and all dates are expressed in terms of a calendar (based on the motions of the Sun and Moon). The importance at this early stage of *astrology*, which was later to become so vital, is unclear, although there is no doubt that the Shang people used other means of divining the future – such as oracles – on a grand scale.

The people of the subsequent Chou Dynasty made regular astronomical observations and astrology became one of the standard methods of divination. It was probably at this period that the Chinese first defined their traditional star groupings, which bear virtually no relation to the familiar constellations of the West. The Chinese divided the sky, as far south as could be observed, into several hundred small "asterisms", or star groups, each containing only a very few stars. Numerous sky maps have survived from later periods and these show little variation over the centuries. Most of them also clearly represent the "Heavenly River" (the Milky Way).

Chinese astronomy was from the earliest times based on the celestial equator rather than, as in the West, the ecliptic – the path of the Sun through the sky. From a very early period (quite when is impossible to say), astronomers attached special importance to 28 asterisms spread out roughly along the celestial equator. These are the so-called "lunar mansions". In later Chinese astronomy these lunar mansions were used to define zones of right ascension (celestial longitude). The boundaries of these zones appear on most maps as radial lines originating at the celestial pole.

The foundations of the highly complex system of astrology which was such a dominant feature at later periods was laid down during the Chou Dynasty (*c.* 1100–481 BC). The Chinese apparently began to believe that a close mutual influence existed between natural events and human affairs. Thus, by observing unusual celestial and other phenomena, they could predict future misfortune or prosperity. I should mention here that at all periods Chinese astronomers were almost exclusively concerned with what one might call portent astrology, relating to the emperor and state. The process of calculating an individual's fate, however, was not indigenous; Chinese horoscopes as such are founded on ideas imported from the West. At the Chou capital, the "Grand Historian" recorded natural and human happenings, performing concurrently the duties of both astronomer and astrologer. Later in the dynasty, when the power of the ruler was much weaker and China was divided into a number of semi-independent states, each state appears to have had its own court astrologer, who made astronomical observations which he used to give advice to the ruler. About this time, the Chinese began to believe that various lunar mansions corresponded to individual states. Thus if a comet or new star appeared in a particular lunar mansion, or a planet passed nearby, this was an omen for the associated state.

The historian Szu-man Ch'ien gives some fascinating insights into political astrology at this time. For instance, in 480 BC Tzu Wei, the court astrologer to Duke Ching of the state of Sung, observed the planet Mars "guarding" Hsin lunar mansion (three stars in Scorpius including Antares). This star group was thought to represent Sung, and the duke was alarmed that misfortune would soon befall him. Tzu Wei accordingly advised him that the impending calamity could be diverted either onto his ministers or the common people, or alternatively the harvest. Ching nobly rejected each of these proposals, whereupon the wily Tzu Wei replied: "Heaven is high above, but it listens to what is said below. Your Highness has made three wise decisions. Mars will change its position." Further observation showed that the planet had moved by 3 degrees. Such was the astrologer's power!

The Warring States Period (481–221 BC) was a time of intense philosophical speculation. Astrologers now found analogies of all kinds between the natural and human worlds. By the Han Dynasty (206 BC to AD 220), which followed closely on the unification of the empire, astrology had practically reached maturity. It was then that a rigorous set of rules for making prognostications was laid down, and these were to remain virtually unchanged and unchallenged throughout subsequent dynasties right down to modern times. On this system, each asterism, and often each single star, represented anything from the emperor or government ministers down to the common people and even everyday objects and needs. Any unusual phenomenon involving a particular asterism or star was a precursor of an event influencing the associated person or object. It is interesting that bad omens were about twice as numerous as good ones! Although rules for making prognostications were remarkably rigid, a considerable length of time – typically about two years – could elapse before the acknowledged fulfilment of the omen. Naturally, this ensured at least a reasonable chance of success.

The more spectacular phenomena, especially eclipses and comets, were particularly serious admonitions. The following imperial decree is a good illustration: "During the night of the 15th of the 7th moon, a flight of meteors occurred in the southern hemisphere; then days later, a comet twice appeared in the northwestern sky.... The empresses dowager now give warning that those portents from Heaven are sent because of serious wrong in our system of government." This might easily have been written during the Han Dynasty, for many similar edicts were written during this period. In fact it dates from AD 1861 and comes from the pen of Tzu Hsi, the empress at the time of the Boxer Rebellion.

As I have already emphasised, the main motive for astronomical observation in China was astrological. There is no doubt that the Chinese regularly watched for solar eclipses to help maintain a reliable calendar. However, the very fact that prognostications accompany numerous records of solar eclipses, and the existence of vast numbers of other types of observations which are of no relevance whatever to the calendar, underline the fundamental importance of astrology. It was during the Han Dynasty that the so-called Astronomical Bureau was founded. This was staffed by civil servants whose function was to make observations of all kinds and where necessary to warn the emperor of impending calamities without delay. By the Ming Dynasty (AD 1368–1644), when the first Jesuit missionaries arrived in China, more than 30 astronomers manned the Imperial Observatory, so that it was probably the most well-staffed observatory in the entire world. However, the fundamental emphasis had not changed over the centuries. The Jesuit Father Sabatino de Ursis, writing from Peking in 1612, gives an intriguing account of the operation of the Bureau as seen through European eyes:

> The kings founded a bureau or special college for this science [astronomy], and its members have no other duty than to calculate eclipses, to make the calendar each year, and to observe the stars, the comets and other prodigious phenomena of the sky, daily and nightly, for the purpose of

advising the king and of declaring whether these are good or evil omens. For others besides the official members of this bureau to do this work is prohibited under grave sanctions. . . .

The element of secrecy that de Ursis noted seems to have been standard at all periods. For instance, during the T'ang Dynasty, an imperial edict forbade the court astronomers to divulge their business even to other civil servants. As a result, persons other than members of the Bureau seldom made observations of importance. Most of the 26 dynastic histories contain an "astronomical treatise", in one or more chapters, and this is essentially a summary of the day-to-day reports of the imperial astronomers. The early discovery of printing by means of wooden blocks – perhaps during the 7th century AD – allowed the various official histories to be printed and reprinted many times and so they are easy to obtain.

The range of data contained in one of the astronomical treatises provides a good example of the diversity of observations made by the imperial astronomers. The *Sungshih* ("History of the Sung Dynasty", covering the period from AD 960–1279), lists as many as 19 different types of observation. These include: solar eclipses, solar "fogs" (sunspots and atmospheric dimming of the Sun); lunar eclipses, lunar "fogs", lunar haloes; conjunctions of the Moon with the five planets (Mercury, Venus, Mars, Jupiter and Saturn); conjunctions of the Moon with asterisms; conjunctions of the five planets with asterisms; sightings of Venus and occasionally Jupiter (!) in the daytime; the five planets in conjunction with one another; sightings of Lao-jen-hsing (the "Old Man Star" – the bright southern star Canopus, which is not well seen in middle northern latitudes); "auspicious stars" (a mixture of comets and novae), "broom stars" (comets); "guest stars" (usually novae or supernovae, along with the occasional comet); "flowing stars" (meteors); "baleful stars" (usually comets); star "fogs" (dimming of certain stars by mist, etc.), and "cloudy vapours" (usually aurorae). This constitutes a catalogue of almost every kind of astronomical observation which can be undertaken with the unaided eye.

Chinese observations which have attracted most attention among present-day astronomers are supernovae, sightings of Halley's comet, sunspots and solar eclipses. The supernovae which appeared in AD 1572 and 1604 were duly reported in the Far East, but were much more accurately observed in Europe by such notable astronomers as Tycho Brahe and Johannes Kepler. (This was still before the introduction of the telescope.) However, for accounts of earlier supernovae (AD 185, 1006, 1054 and 1181), we have to rely almost entirely on the records of China and Japan. Again, virtually all of the pre-telescopic sightings of sunspots were made in China and Korea, the earliest in 28 BC. An important question concerns the reliability of observations of this kind, which cannot be checked by back-calculation. This can best be answered by looking at the various types of observation which *can* be so verified.

Table 1 Supernovae observed in China

Date	*Constellation*	*Visibility*	*Status*	*Radio remnant*
AD 185	Centaurus	20 months	probable	G315.4 − 2.3
393	Scorpius	8 months	possible	G348.5 + 0.1 (?)
1006	Lupus	several years	established	G327.6 + 14.5
1054	Taurus	22 months	established	Crab Nebula
1181	Cassiopeia	6 months	probable	3C 58
1572	Cassiopeia	16 months	established	3C 10
1604	Ophiuchus	12 months	established	3C 358

I have analysed many reports of lunar and planetary conjunctions at various periods in Chinese history. Almost all prove to be thoroughly reliable. Although *measurements* of any kind are rare, dates are almost invariably exact to the very day. This gives me considerable confidence in the other types of observation which have to be accepted on trust, such as those of sunspots and new stars. A very impressive testimony to the reliability of the Chinese astronomers concerns their sightings of Halley's comet, which is due back in 1986. The Chinese reported almost every apparition of this comet since 12 BC.

The only type of observation for which there is any significant evidence of deliberate fabrication is of eclipses of the Sun and Moon. A substantial proportion of all eclipse reports in Chinese history relate to predictions rather than observations. In some cases the prediction is only thinly disguised. There are many such statements as: "The Sun should have been eclipsed but on account of thick clouds [or rain] it was not visible." Prediction of solar eclipses was generally so bad in China that it is easy to detect false sightings. Especially in the earlier dynasties, most of the calculated eclipses missed the Earth completely and scarcely any were visible in China! However, with lunar eclipses it is not always possible to distinguish between calculation and observation. The prediction of eclipses seems to have become something of a tradition at quite an early period in Chinese history. Fortunately this did not extend to other phenomena, and as a general rule Chinese observations are of the highest reliability, and represent a body of early astronomical data without parallel anywhere else in the world.

4

Ancient eclipses cast light on modern science

RICHARD STEPHENSON

22 February 1979 and 19 August 1982

A total eclipse of the Sun is so awesome that accounts survive from the earliest times. The mere knowledge that eclipses were seen at particular places on certain dates is surprisingly powerful information. Modern astronomers can use the data to investigate the Earth's rotation, the Moon's motion, the strength of gravitation, and changes deep within our planet.

Interest in historically observed eclipses goes back at least as far as the time of Claudius Ptolemy. About AD 150, this noted astronomer of Alexandria investigated a number of eclipses of the Moon seen in Babylon long before his own time. He used these observations, which are reported in his monumental treatise *Mathematike Syntaxis* (later known as the *Almagest*), in his theory of the Moon's motion. The Babylonians appear to have kept systematic records of all kinds of astronomical phenomena from as far back as about 750 BC. Whether Ptolemy had access to much of these data is impossible to say since he only cites lunar eclipses. However, what is particularly impressive is that all the dates of the eclipses to which he refers are precisely recorded, the earliest corresponding to 19 March, 721 BC. Despite much current research into original Babylonian astronomical texts, notably by Abraham J. Sachs of Brown University, Rhode Island, this eclipse in 721 BC remains the most ancient reliable Babylonian astronomical observation of any kind.

Investigation of historical eclipses, mainly of the Sun rather than the Moon, still continues and remains much concerned with the motion of the Moon. As early as AD 1695, analysis of such data led Edmond Halley (of comet fame) to suspect that the mean motion of the Moon was being slowly accelerated, an idea amply confirmed by another Englishman, Richard Dunthorne, half a century later. This acceleration remained a unique problem until the beginning of the present century. In 1905, analysis of a number of ancient eclipses led Philip Cowell of the Royal Greenwich Observatory to go one step further. He suggested that the Sun also had a small acceleration, equivalent to an increase in speed of the Earth in its orbit as seen from the standpoint of the Sun. The idea was hotly disputed since there appeared to be no good physical reason for Cowell's contention.

It was not until 1939 that the problem was finally solved by Sir Harold Spencer Jones, then Astronomer Royal. From telescopic observations made since the 17th century he demonstrated that the Sun has no real acceleration; instead the Earth's poor time-keeping is the culprit! The rotation of the Earth is not perfectly uniform but is slowing down. What appeared to Cowell to be a speeding up of the Sun was nothing more than a gradual lengthening of the standard unit of time – the mean solar day. Spencer Jones was able to support his conclusion by studies on the motions of the Moon and Mercury.

The gradual lengthening of the day is now known to be mainly due to the tides produced by the Moon and the Sun in the oceans and body of the Earth. On account of friction, the Earth's energy of rotation is being very slowly dissipated. Direct evidence of the effects of tidal friction is seen in the fact that the Moon always turns the same face towards the Earth. Here the enormous tides produced by the Earth in the solid body of the Moon are responsible. Similarly the two small satellites of Mars also keep the same face towards the planet.

For very good reasons, astronomers no longer measure time by the rotation of the Earth. Instead, they use a dynamical time defined by the motion of the Sun. As we might expect, the apparent acceleration of the Sun and planets disappears in this system. However, the Moon is instead observed to be slowing down. The explanation here is directly linked to the Earth's loss of speed by tidal friction. Conservation of angular momentum is a well-known principle in physics and this applies to the Earth–Moon system. As the lunar tides retard the Earth's rotation, angular momentum is transferred from the Earth to the Moon. Now while not immediately obvious, it is nevertheless easy to prove that as the Moon, or any other orbiting body, gains angular momentum it slows down and the size of its orbit increases: the Moon is gradually receding from the Earth.

Ancient eclipses have thus played a part in our understanding of two closely linked problems in astronomy: changes in the motion of the Moon and in the rotation of the Earth. Even now the importance of these eclipses is undiminished because of their extreme age. Both the error in the Moon's position and the clock error caused by the lengthening of the day increase as the square of the time. In the 2700 years or so covered by almost all usable observations the Moon has receded from the Earth by only about 100 metres (its diameter is 3500 kilometres), and the length of the mean solar day has increased by approximately 0.05 seconds. Yet the cumulative effects of these tiny changes are not only detectable but measurable. Around 700 BC the Moon was as much as 3 degrees behind its expected position. Equally, a hypothetical clock set in this remote epoch would be now about six hours fast on the Earth. Even primitive observations can reveal these large discrepancies.

It is when we try to separate the effects caused by the deceleration of the Moon and the slowing rotation of the Earth that we encounter real difficulties. Almost all ancient observations which are of any value in the present type of study involve the Moon, either in the form of eclipses or occultations. It takes many observations to separate satisfactorily the two effects just mentioned. As might be expected, ancient attempts at measuring time were generally far from reliable. For this reason much research has been done with untimed, but accurately dated observations. These all relate to central (or near-central) total and annular solar eclipses. Because of the rather special geometry, times are unnecessary here, provided the observer has given a fairly careful description of what he saw.

An eclipse of the Sun occurs whenever the Moon passes directly between the Sun and the Earth so that its shadow falls upon the terrestrial surface. The Moon is a fair-sized object (diameter 0.27 times that of the Earth) so that its shadow sweeps over a significant area. However, because of a remarkable chance coincidence, the zone of total shadow on the Earth in which no part of the Sun is visible is extremely narrow, never more than a few hundred kilometres wide, and may be nonexistent. The explanation is that at the Moon's average distance from the Earth it appears to have almost exactly the same angular size as the Sun. However, because of its rather elliptical orbit, the Moon is often slightly too far away to completely hide the Sun; instead a brilliant ring of unobscured sunlight is visible at central eclipse. These annular eclipses are roughly as frequent as total eclipses. In most years one of each type occurs somewhere on the Earth's surface. However, at any given place there are typically three total eclipses and four annular eclipses visible every 1000 years. The last eclipse of the Sun which was total at London occurred in AD 1715.

Annular eclipses seem to attract little attention, but, on the contrary, total eclipses rank as perhaps nature's most awe-inspiring and breath-taking spectacle. This is illustrated by the large numbers of total eclipses recorded in history (some 50); by contrast, annular eclipses are almost never recorded. Rather than attempt to describe the various phenomena which accompany total solar eclipse, let us leave the description to an eyewitness who lived fully 1000 years ago. Leo Diaconus, who wrote a history of Byzantium covering part of his own lifetime, includes the following account of the total solar eclipse of 22 December 968.

> While these things were being done by the Emperor in Syria, it happened that there was an eclipse about the time of the winter solstice of the Sun.... And the eclipse was of the following form. December was setting in motion its twenty-second day, and at the fourth hour of the day, while the weather remained calm and clear, darkness covered the Earth and all the brightest stars shone forth. And it was possible to see the disc of the sun, dull and unlit, and a dim and feeble glow like a narrow band shining in a circle around the edge of the disc. Gradually the Sun passing by the Moon (for the Moon was seen to be obstructing it in a straight line) sent forth its own rays and again filled the Earth with light.... At that time I myself too used to visit Byzantium in pursuit of general education.

The last sentence makes it somewhat doubtful whether Leo himself saw the eclipse. However, he tells us at the beginning of his history that what he has to relate is based on either his own experience or the eyewitness accounts of others whom he personally consulted. It is worth remarking that Leo is one of the few medieval European writers to reveal a clear understanding of the nature of a solar eclipse.

Notice that Leo alludes to the time of day, but only very crudely; this is fairly common among the ancient and medieval reports of eclipses. However, it is fully evident from what he tells us that the city of Byzantium must have been within the narrow zone of totality. The solar photosphere has a fairly sharp edge, and although the limb of the Moon is rough with numerous mountains, it is relatively easy for an observer equipped with only the unaided eye to decide just when the eclipse is total.

Most of the 50 or so reliable observations of total eclipses which have come down to us from ancient and medieval times are recorded in historical works, rather than treatises on astronomy. Nevertheless, they all have three fundamental properties in common: the complete disappearance of the Sun is clearly described; the date is given or can be readily established; and the place of observation is either implicitly stated or can be confidently inferred.

The very earliest record comes from the ancient seaport of Ugarit, whose ruins are in modern Syria. The text is inscribed on a clay tablet in a cuneiform alphabetic script. It can be dated, partly as a result of astronomical calculation, to 3 May 1375 BC. Virtually all the other ancient observations (before about AD 500) are from China, and are probably the work of professional astronomers employed by the ruler at the capital. The earliest of these dates from 709 BC. Perhaps surprisingly, most of the medieval accounts are from European monastic chronicles. These mainly detail local events, and frequently report the more spectacular astronomical phenomena such as bright comets and large solar eclipses.

To illustrate the use of untimed observations of totality, we can refer to what is probably the most reliable account of an eclipse before AD 1500. On a fragmentary tablet, which is part of an astronomical diary for the year 175 of the Seleucid era (137–136 BC), Abraham Sachs discovered a fascinating account of a total eclipse of the Sun. The tablet was originally found in the ruins of the city of Babylon and is now in the British Museum. Although there is no direct reference to the degree of obscuration of the Sun, it is certain that a total eclipse is referred to. Fortunately a separate

An unusually well-preserved clay tablet records in cuneiform "wedge" characters Babylonian predictions for celestial phenomena in the year 118 BC. It also details observations of the total solar eclipse of 15 April, 136 BC "a few minutes before sunset"

tablet describing the same event mentions specifically that the eclipse was total.

If we calculate the position of the track of totality in 136 BC, using a plausible value for the Moon's deceleration, but make no allowance for the changing length of day, the eclipse track does not come within 3000 kilometres of Babylon (Figure 1). The records seem thoroughly reliable, so we must infer that our calculations are based on an incorrect assumption; the Earth's rotation *is* slowing down, and at such a rate that the accumulated clock error since 136 BC amounts to nearly five hours.

Comparison of modern calculations with Babylonian eclipse records does in fact give an excellent figure for the average rate of lengthening of the day over the past 2500 years. It works out 1.78 milliseconds per century. The uncertainty in this figure is about 0.14 milliseconds, and it arises not so much from inaccuracies in the Babylonian measurements but more from uncertainties in the modern estimates of the rate of expansion of the Moon's orbit. Because the visibility of an eclipse depends on the Moon's motion as well as the Earth's rotation, we must determine this quantity accurately. Reliable results should be available over the next few years

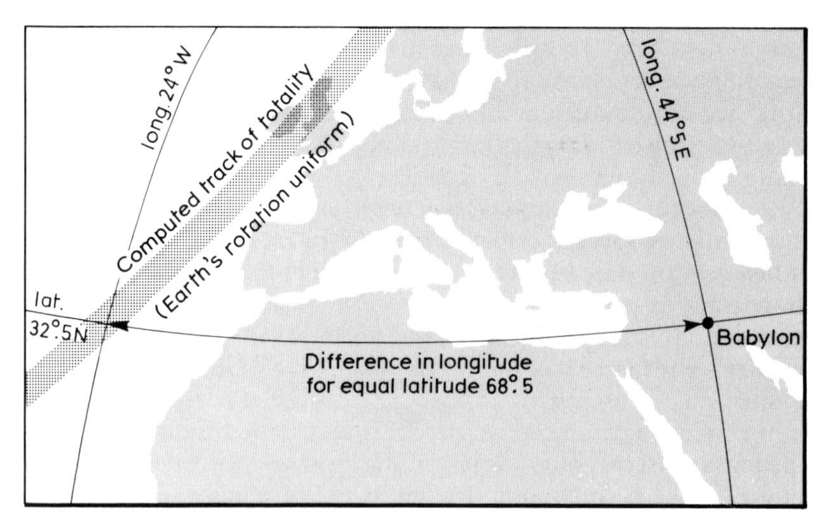

Figure 1 *If the Earth were not slowing down, the total solar eclipse in 136 BC would only have been seen from the Atlantic. Since cuneiform tablets record that it was seen in Babylon, the difference in latitude, marked here, provides an accurate way of calculating the rate at which the Earth's rotation has been slowing over the past two millennia*

from accurate range-finding which involves reflecting beams off mirrors left on the Moon by the Apollo astronauts and unmanned Russian probes. Only then will the full potential of the ancient data be realised!

The data already available has been used to settle a modern cosmological problem, the possible change in *G*, the universal "constant" of gravitation. In the 1970s, Thomas Van Flandern of the US Naval Observatory, Washington, presented observational evidence in favour of a diminishing *G*. Since 1955 astronomical observations have been made on Atomic Time, as well as Dynamical Time which is defined by the motion of the Sun. If *G* is decreasing, the planetary orbits and the lunar orbit around the Earth would be expanding. This would not be detectable on Dynamical Time because this timescale is defined in terms of the planetary motions. However, Atomic Time is an independent time system and the effect would be evident. The lunar laser ranging measurements are effectively on an atomic timescale, whilst the results from the ancient and medieval data and also from the transits of Mercury between 1677 and 1973 are on Dynamical Time. The close agreement between the two pairs of results is strong evidence in favour of the constancy of *G*. It would appear that the rate of change of *G* – if it is changing at all – is less than about 1 part in 10^{11} per year. This is the best observational evidence so far obtained; it is an order of magnitude more accurate than that currently available from planetary radar ranging. It also rules out Van Flandern's original claim that *G* is changing by 10 parts in 10^{11} per year.

Perhaps the most important conclusion of the investigation of Babylonian eclipses is evidence that the Earth is not slowing down as fast as would be expected. The argument goes as follows: if the change in the speed of the Earth's rotation was due only to the tides, then the day should lengthen by more than 1.78 milliseconds per century – the figure would be more like 2.5.

This slight discrepancy must be accounted for, geophysicists think, by changes in the way mass is distributed inside the Earth. The mass must be gradually becoming more concentrated nearer the centre. To keep the books straight as far as angular momentum is concerned, the Earth must speed up to compensate, and this partially offsets the tidal slowing. So as a result of their painstaking observations, the ancient Babylonian astronomers have unknowingly made an important contribution to modern geophysics as well as astronomy.

PART THREE

Radio Astronomy Pioneers

Radio astronomy is one of the newest sciences. It was promoted from the speculative to the practical largely by discoveries which radar specialists made during the Second World War.

Since then its progress has been swift and astonishing. It is enabling scientists to probe astronomical mysteries which were previously veiled by the limitations of optical telescopes. It is adding immensely to our knowledge of the solar system and regions far beyond.

In Britain the pioneer work on radio astronomy has been notably bold and successful. This work will be further aided by the steerable radio telescope which has been erected for the University of Manchester at Jodrell Bank, and by the two radio interferometers nearing completion in the new Mullard Radio Astronomy Observatory near Cambridge.

So ran a lead article in *New Scientist*, in August 1957. Interest in the new science ran so high that the magazine launched its first-ever series, eight consecutive articles as *A Guide to Radio Astronomy*. Pioneer radio astronomers had already detected radio waves from the remains of supernovae and from other galaxies – many of them very distant – and mapped the distribution of hydrogen gas in the Milky Way, analysing its velocity structure to reveal our Galaxy's spiral arms.

The optimism was entirely justified. During the next two decades, radio astronomers were to discover the regularly "ticking" pulsars, complex molecules in space, and, beyond our Galaxy, the immensely powerful quasars and even echoes of radiation from the big bang in which the Universe began.

Radio astronomy became the first of the "new astronomies" by an accident of nature. Apart from light, radio waves are the only radiation from space which can penetrate Earth's atmosphere. The radio "window" stretches from wavelengths of around 1 centimetre to about 30 metres. Hence radio telescopes can be built and operated conveniently on the ground, unlike the high-flying telescopes needed

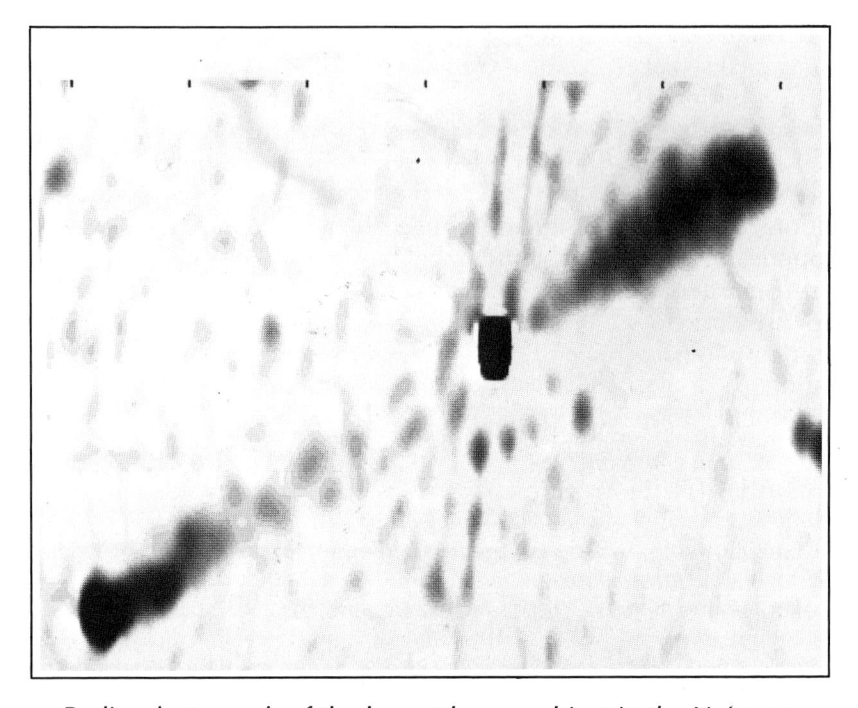

Radio photograph of the largest known object in the Universe, 3C 236. The huge lobes, detectable only by their radio emission, stretch out over 20 million light years. They have been produced by energy beamed out from a giant elliptical galaxy coinciding with the central rectangle

in, for example, X-ray astronomy. The early radio astronomers started with the long wavelengths, which were easier to cope with, but improvements in electronics now mean that the entire range is accessible, right down to the arbitrary limit between radio and infrared at 1 millimetre (although water vapour in the atmosphere makes observations at wavelengths shorter than 3 centimetres a chancy business from observatories at sea level).

A radio telescope is essentially a highly sophisticated radio set, no different from the domestic kind. It has an antenna (or aerial) to pick up the radio waves; an amplifier to boost the very weak signals; and an output stage. Only in the output stage does the telescope differ significantly from the radio set, for radio astronomers are not interested in "listening in" to radio waves from the sky. Astronomers have other ways of displaying the information in the waves to investigate exactly what is producing radio emission.

The antenna system of a radio telescope is what draws a passer-by's attention, especially when it involves hundreds of tonnes of ironwork. But it is possible to make a radio telescope with the simplest antenna of all. A dipole aerial is a straight length of wire, broken in the middle, and with the signal produced by an incoming radio wave tapped off as the voltage difference between the two halves of the wire. Such aerials are sometimes used for ordinary FM radio sets, and they are directional: swivel a dipole around and it will respond most strongly when it is at right angles to the direction of the broadcasting station.

To locate sources in the sky, radio astronomers seek far better directionality. It turns out that improving directionality goes hand in hand with solving another problem of radio astronomy – that radio waves from space are incredibly weak. In turning over a page of this book, you are expending more energy than has been picked up by all the radio telescopes in the world over the past 50 years! So a radio telescope needs to scoop up radio waves from as large an area as possible. And, neatly enough, a large telescope also gives great directionality.

The big dish fulfils the role of a large telescope admirably. The curved bowl intercepts radio waves over a large area, and reflects them to a focus above its centre. Here a dipole antenna converts the concentrated radio waves into an electrical signal. Only radio waves coming from the part of the sky directly in front of the dish are focused onto the dipole. By scanning the telescope back and forth, up and down, an astronomer can sweep out a path across the sky, locating radio sources and building up a picture of the radio sky.

In practice, such a radio telescope is sensitive not only to the exact point in front of it, but to a small region of sky around it, picked out by what is known as the telescope beam. So if two sources lie close together in the sky, separated by much less than the width of the beam, the telescope will "see" them as just one source. Reducing the size of the beam (which is equivalent to improving the directionality) is one of radio astronomy's continuing activities.

The beam size is determined by the area of the dish (relative to the wavelength of the radio waves detected). A larger dish gives a smaller beam, and hence a better ability to resolve the fine detail in a radio source. This is the other reason why radio astronomers have constructed ever-larger dishes, or have used techniques that synthesise a large dish by using smaller aerials connected electronically.

Much of the glamour has faded from radio astronomy with the passage of time. It has also been eclipsed in the public eye by the growth of the other new astronomies, which were largely inspired in fact by the successes of this pioneer science. The now familiar radio dishes lack the "pop appeal" of sophisticated satellites operated by remote control, opening new frontiers from the infrared wavelengths right down to gamma rays. Radio astronomy's role has to a large extent consolidated. Astronomers now only rarely construct radio telescopes for one specific investigation. Large expensive telescopes are built as general-purpose instruments, with many different astronomers using the telescope for their own investigations. These will often be part of a research programme that involves observations at optical wavelengths too; and sometimes the project will require measurements over the whole range of the electromagnetic spectrum.

But its changing public image does not imply that radio astronomy is any the less important; indeed the establishment of radio observatories means that astronomers now include radio observations as naturally in their investigations as optical observations. Radio astronomy still has a pioneering role. Its spearheading of research is particularly marked in two areas. One is the study of molecules in space, which by their emissions of specific wavelengths produce spectral lines. Although the first lines studied were at wavelengths of a few centimetres, molecular line astronomers are now developing techniques to study the huge number of lines at wavelengths of a millimetre or less, in the borderline "submillimetre–millimetre wave" part of the spectrum. This topic is covered more fully in Part Five.

The second main field of advance is the use of paired telescopes as interferometers, to give very high resolution. The major instrument at many radio astronomy observatories is a large array of radio telescopes, which astronomers can use routinely to observe finer details than any optical telescope can "see". Telescopes separated by even greater distances – hundreds or thousands of kilometres – can show details as small as 0.001 arcsecond, one-thousandth the size of the images from optical telescopes, or any other type of telescope. So far this technique of Very Long Baseline Interferometry (VLBI) is used only when several telescopes can be coordinated, often by arrangements that must be international. Plans are, however, afoot for purpose-built networks in the United States, Canada and Australia that will reveal this kind of detail on a routine basis. Observations so far have led to one of the most exciting and perplexing discoveries of

Some of the 25-metre dishes making up the Very Large Array near Socorro, New Mexico. There are 27 dishes in the Y-shaped array, with each "arm" of the Y some 20 kilometres long

the 1970s: in the centres of some quasars, clouds of radio-emitting gas are apparently moving apart at velocities several times faster than the speed of light!

The selection of articles presented here mirrors the development of the science that pioneered the new astronomical revolution. As a result, some radio telescopes have received less space than they deserve. The largest single-dish radio telescope, for example, is of a unique design which lies outside the mainstream of development. This 305-metre bowl of wire mesh, hung in a hollow in the limestone hills near Arecibo in Puerto Rico, cannot be tipped up, and merely watches what passes through the region more or less overhead. Its large area, however, makes it very sensitive, and ideal for studying pulsars. The Arecibo telescope picked up the first radio-emitting pulsar in a double star system, a system whose properties have been used to test the general theory of relativity.

The technique of interferometry developed at Cambridge, called Earth-rotation synthesis, led from the One-Mile Telescope to the higher-resolution Five-Kilometre Telescope. The latter's eight dishes (along the old line of the Oxford–Cambridge railway) form 16 pairs of telescopes simultaneously, synthesising a dish 5 kilometres in diameter. Dutch radio astronomers have constructed a similar array at Westerbork, containing 14 individual telescopes. The technique has culminated in America's Very Large Array, a set of 27 medium-sized (25-metre) radio dishes which can be moved on a Y-shaped system of railway tracks, to synthesise a single dish over 20 kilometres in diameter.

The Very Large Array and Britain's Merlin array – centred on Jodrell Bank – rank as the world's leading multipurpose radio telescopes at present. The account of Merlin (p. 75) not only covers the specific instrument, but indicates the state and status of radio astronomy today. It epitomises the strong roots of the science in the inclusion of the 76-metre dish at Jodrell Bank; it is used as an observatory, in particular as a partner to optical studies of quasars; and it is pioneering the dissection of natural masers in our Galaxy, and the past history of radio galaxies and quasars far out in space.

5

Birth of a new science

GRAHAM SMITH

15 August, 1957

Radio astronomy gave rise to several important discoveries even before the construction of the major telescopes at Jodrell Bank and Cambridge.

The first suggestion that there might be radio waves from outside the Earth came from Thomas Edison. In 1890 he suggested to the principal of Lick Observatory that "along with the electromagnetic disturbances we receive from the Sun – as light and heat – it is not unreasonable to suppose that there will be disturbances of much longer wavelength".

The huge loops of wire which Edison proposed for detecting these radio waves were quite unsuited for the short-wavelength radiation which is now regularly observed from the Sun, and the experiment, if indeed it was completed, must have failed. In Germany, France and England similar experiments were made at the beginning of the century, using various arrangements of wires rather like the aerials used in early radio communication experiments. Sir Oliver Lodge used a sensitive coherer as detector, but the only signals he received were from the sparking of Liverpool trams.

These workers lacked two essentials: sensitive receivers and directive aerials. Both were available in 1932 to Karl Jansky, an American engineer who was investigating background noise levels for short-wave radio communications. He found that the background radiation was not constant throughout the day, but showed a regular variation of intensity, reaching a maximum when the Milky Way was overhead. His observations showed clearly that he was receiving radio waves from outside the Earth: it is these radio waves that take the place of light waves in radio astronomy.

It is surprising that the only successful follow-up of Jansky's observations before the Second World War was by an amateur, Grote Reber, who built an aerial 9 metres in diameter in his back garden. Reber was able with this reflector to measure radio waves from the Galaxy at wavelengths as short as ½ metre, and he carried out surveys of the distribution of emission over the sky which have only recently been improved upon.

As we have seen, the earliest results in radio astronomy came from the United States; but during the war the foundations of great activity in the subject were laid both in this country and in Australia. J. S. Hey, working in the Army Operational Research Group on problems of radar, was responsible for three major discoveries which have each led to their own branch of radio astronomy. First, he showed that the new radar equipment could receive radio waves emitted from sunspots; secondly, that radar sets could obtain echoes from the trials of meteors; and thirdly, that there was a fluctuating source of noise in the constellation of Cygnus, which he decided correctly must come from a point rather than a large area of the sky. This was the first "radio star" to be discovered. It was found to fluctuate or "twinkle" just like an ordinary star – indeed, it was later shown that the cause of twinkling was the same in each case and that the radio stars' fluctuations were due to irregularities in the Earth's ionosphere which changed the intensity of radio waves from moment to moment.

A fourth major discovery came from experiments made in America and Holland on radiation from the clouds of hydrogen between the stars. The radiation we have so far discussed comes in on a wide range of wavelengths (though with greater intensity at longer wavelengths), but from the predictions of Van de Hulst in Holland it appeared that neutral hydrogen would radiate at the one special wavelength of 21 centimetres, and that valuable information about the structure of the Galaxy might be obtained from this radiation. The confirmation of this prediction is one of the most striking successes of radio astronomy.

The design of aerials to further these observations must take into consideration two important factors – the difficulty of attaining sufficient resolving power, and the weakness of the signals. The resolving power defines the accuracy with which the position of a radio star can be determined, and the ability to discriminate between adjacent stars. It is related to the ratio of the size of the instrument to the wavelength, and since radio wavelengths are about 10 million times those of visible light, a radio telescope with a

Early stages in the construction of the world's first large fully steerable radio telescope, the 76-metre dish at Jodrell Bank

discrimination only as good as a human eye would have to extend for 10 miles (16 kilometres).

The signals produced by the faintest sources so far detected have a field strength 10 000 times smaller than that received by a typical television aerial from the BBC. Since there are definite limits to the sensitivity possible with a radio receiver, set by the noise generated by the receiver itself, the only way of detecting the weaker sources is to collect the incoming radiation over a big area, or in other words to use large aerial arrays.

On high frequencies, and particularly for the hydrogen line radiation, these two factors both lead to the paraboloidal aerial, of which type the new 250foot (76-metre) diameter reflector of the University of Manchester at Jodrell Bank is the prime example.

6

The giant at Jodrell Bank

NIGEL CALDER

3 October 1957

The huge radio telescope at Jodrell Bank was a modern wonder of the world when completed in 1957. It involved engineering of a complexity never before attempted. Later telescopes benefited considerably from the lessons learned at Jodrell Bank – including the unexpectedly high cost of £850 000. This article was published the day before the Space Age began, with the launch of Sputnik 1. The subsequent story of spacecraft-tracking at Jodrell Bank made the telescope world famous, and the publicity prompted Lord Nuffield to clear the debt that had hung over the telescope since its opening.

Travellers by train from Manchester to Crewe cannot fail to be impressed by the new 250-foot (76-metre) radio telescope which stands above the Cheshire countryside at Jodrell Bank. The reflecting dish, large enough to seat 10 000 people, is suspended 60 metres up; when it turns and tilts to scan the sky or follow a star, the 2000-tonne structure can move with the delicacy of the hour hand of a clock. Bernard Lovell, at Manchester, who camped gipsy-wise on the old Botanical Station at Jodrell and became the first professor of radio astronomy, is a bold, imaginative spirit. He foresaw 10 years ago the potentialities of such a huge steerable dish, and subsequent discoveries have confirmed his judgement.

A radio astronomer wants the biggest aerial he can afford, so that he can detect the feeblest radio waves and can measure precisely the

---

(Opposite) *Work is completed inside the Jodrell Bank dish. The original steel plates, which focused radio waves onto the aerial* (centre), *were fixed to an accuracy of half a centimetre. The surface has since been improved considerably*

direction from which they are coming. In a paraboloidal dish reflector of the type used at Jodrell, the aerial proper is carried on a mast sticking out from the centre of the dish, and if it is pointed at a radio star the waves are reflected from the metal surface and focused onto the aerial. But the dish is not indefinitely large and, at the rim, the waves are not reflected geometrically. Radio waves from parts of the sky offset from the axis of the telescope can therefore reach the aerial by reflection at the rim. The strength of rim reflections compared with those from the whole area inside the dish limits the power of the telescope to distinguish or resolve two objects close together. The shorter the wavelength, or the larger the dish, the better is the resolving power.

So Lovell's requirements can be easily expressed: a scaled-up version of a radar dish aerial, 76 metres in diameter and accurately parabolic in cross-section. It should be able to point in any direction – including straight down (reversing the stresses in many of the members) so that the radio astronomers may rapidly change the aerial on the 19-metre mast. And in order that the shortest radio waves can be detected and amplified as close to the aerial as possible, there has to be a laboratory swinging behind the centre of the dish.

H. C. Husband, a consulting engineer from Sheffield, designed the Jodrell telescope. He had to take account of the capabilities of the contractors and it was a typical job for none of them. The steel members were prepared at Scunthorpe and every component, down to the last bolt, has been made in Britain – mostly in Yorkshire. Where precision was demanded it has been given and to spare.

The telescope has the biggest dish ever, weighing 800 tonnes, and the parabolic shape must be preserved in spite of tendencies to distort as it is tilted. A high wind exerts a force of several hundred tonnes on the telescope, yet experiments must go on in winds of up to 70 kilometres per hour and the structure must be safe in a hurricane. As a great scientific instrument, it must perform its rock 'n' roll with uncanny smoothness under the guidance of electronic computers 180 metres away in the control building, and never depart more than $^1/_{12}$ degree from the prescribed direction.

In his first design, of 1953, Husband proposed to make the dish of stretched copper netting. The final design, reached in 1955, achieves a more accurate parabolic profile with bent steel plates.

The construction began with 150 piles driven through the sand and clay of Jodrell to find a foothold on the underlying red marl. In the muddy winter of 1952–53, 4000 tonnes of concrete were poured into the foundations. Then a circular railway was laid, to an accuracy of 1½ millimetres.

A 90-metre bridge of girders turns on a pivot at the centre and on six precision bogies at each end designed to reduce friction to a minimum. Above the bogies rise the two 56-metre towers, carrying the trunnions on which the dish itself rotates. The framework for the dish consists of a central hub and 16 radial wedge-shaped sections – "the slices of the orange" – and as the dish took shape each section was raised in a cradle to the top of the scaffolding and there joined on. Around the inside of the dish are rings of purlins, set to an accuracy of 6 millimetres, and it is on these that the 7000 steel plates of the reflector have been welded.

At a late stage, Husband decided to add a stabiliser to prevent the telescope fluttering in the wind. Hence the great hoop on the back of the bowl (he calls it the bicycle wheel) which runs on rubber wheels carried by two upturned chassis at the base of the telescope. By applying hydraulic brakes to the wheels, movements due to wind can be checked, by applying an upthrust to the stabiliser fine adjustments to the shape of the telescope can be made.

A scale model was thoroughly tested in the wind tunnel at the National Physical Laboratory, and full-scale wind tests are being carried out on the completed telescope to provide data for the radio astronomers. Most of the power of the driving system is held in reserve for "parking" the telescope in a safe position in a high wind.

The driving motors have a great range of speed: from 10 to 1500 revolutions per minute. At full speed the telescope can swing round the compass in 18 minutes and the dish can loop the loop in 15 minutes. The rollers of the railway bogies run on double tracks, and they are slightly conical to take account of difference in radius of the tracks. The tilting power of the motors at the tops of the towers is transmitted by gun-turret racks taken from the scrapped battleships *Royal Sovereign* and *Revenge*, and the overall gear reduction is 21 500 to 1.

The actual orientation of the dish is measured automatically and compared with signals from the control room showing the desired orientation. If any discrepancy exists, the output of the electric generators at the base of the telescope is immediately adjusted to correct the error.

This, then, is the biggest fully steerable telescope in the world. It began working in August, and reports of the first observations are awaited eagerly.

The world's largest fully steerable radio telescope, at Effelsberg, near Bonn, was completed in 1971. This 100-metre dish is operated by the Max Planck Institut für Radioastronomie, in Bonn

7

Germany builds the largest radio dish

PETER STUBBS
13 May 1971

After 14 years, Jodrell Bank's 76-metre telescope lost its title as the "world's largest fully steerable dish" to a new German telescope, 100 metres in diameter. Such is the difficulty, and cost, of constructing large dishes that the Effelsberg telescope retains the title to the present day – and for the foreseeable future.

A new era in European radio astronomy is dawning. This week the West German minister for education and science, Professor Hans Leussink, in company with the president of Nordrhein-Westfalia, officially opened the impressive 100-metre fully steerable radio telescope in the Effelsberg a short distance southwest of Bonn. Since it began to operate, less than three weeks ago, this high-precision instrument has usurped the position of pre-eminence that Jodrell Bank has held with its Mark I radio dish since 1957. For the time being West Germany will have the biggest fully steerable dish.

But, in fact, cooperation rather than competition will typify the coming activities in European radio astronomy. The development of the powerful technique of long-baseline interferometry, permitting much higher resolution of radio sources than hitherto, has already shown its paces in a wide variety of international links between radio astronomy observatories. And the possession by Europe of two, and possibly eventually more than two, major steerable radio telescopes opens up exciting opportunities in this direction.

However, considered on its own, the West German instrument is a remarkable piece of design, rationalising the whole concept of radio telescopes in the realisation of a big dish able to probe the radio sky down to wavelengths of only 1 centimetre. The requisite figuring of the dish is achieved by using a so-called "passive" homology design – when the dish tilts the distortions are such that the paraboloid stays parabolic and only the focus and depth of focus change. These factors are easily taken care of by modern computer control. The Effelsberg dish is some four times more precise than the 64-metre Parkes instrument in New South Wales and the Greenbank, Virginia, 43-metre dish. The advantage of being able to measure at centimetre wavelengths is that it enables radio astronomers to study the microwave emission lines of the many molecules now coming to light in interstellar space.

The telescope itself is sited in the bottom of a bowl-shaped valley which screens it from the disturbing effects of commercial microwave links and radar transmissions. A series of tetrahedrons supports the big dish. The 100-metre reflector itself rests upon an umbrella-like structure of tubular struts which is effectively constrained at only one point.

The surface of the dish consists of a number of circular regions. Out to a radius of 30 metres is the most accurately figured part (good to 0.2 millimetre), consisting of a 10-centimetre thick aluminium honeycomb sandwich faced on either side with 2-millimetre thick formed sheets of aluminium. Between 30 and 42.5 metres radius the surface is a simple 2-millimetre aluminium facing on an aluminium support. The plated region is perforated round its outer 2.5 metres to allow a smooth transition in aerodynamic characteristics to the outermost 7.5 metres which are of 6 millimetres square stainless-steel mesh. The instrument is designed to function in winds up to 30 kilometres per hour.

The full 100-metre dish is expected to operate down to 5 centimetres wavelength, the 85-metre aperture to 2 centimetres, and the inner aperture to 1 centimetre. Observations can be made at both the prime focus and a secondary focus. Computer control for the telescope and also for initial processing of data will be performed by a Ferranti Argus 500 computer.

Design of the DM 34 million instrument began in 1966 and building commenced in 1968, the same year as the formation of a Max-Planck Institut für Radioastronomie, directed by Otto Hachenberg, Peter Mezger and Richard Wielebinski. The bulk of the money was provided by the Volkswagen Foundation, with additions from the West German science ministry and the state of Nordrhein-Westfalia. Krupps produced the successful engineering design but the instrument has been built without a private contractor, Professor Hachenberg describing himself as clerk of works!

As well as interstellar chemistry and space research, the Effelsberg

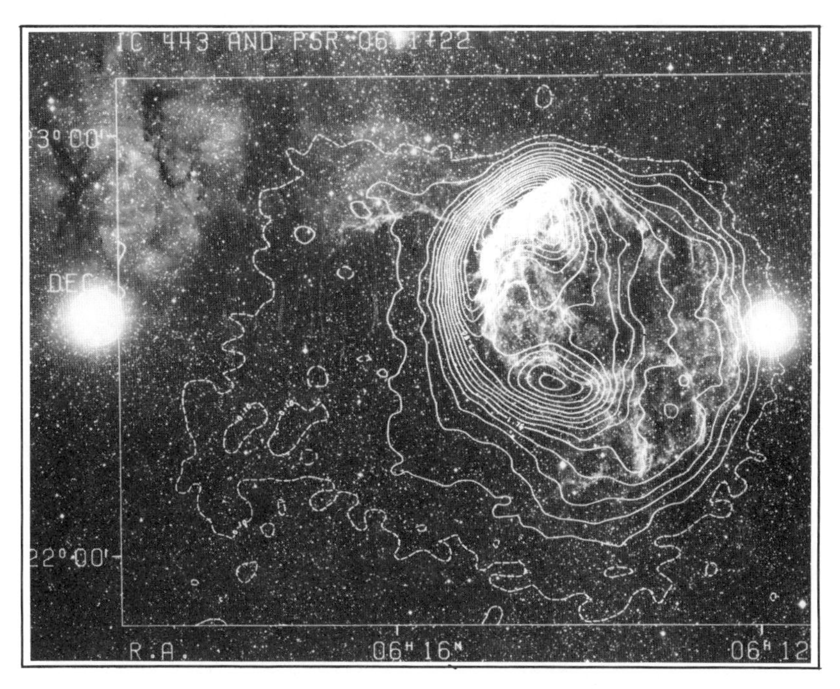

Observations by the Effelsberg telescope at a wavelength of 11 centimetres are shown as a contour map overlaid on an optical photograph. The powerful radio source, and associated optical nebula IC 443, are the remains of a supernova which exploded 15 000 years ago

telescope will carry out work on both the galactic continuum emission and extragalactic objects of various kinds. Its design is particularly suitable for studying pulsars. But for unravelling the structures of quasars, interferometry with very long baselines will be the order of the day. Such studies demand simultaneous observations by stations several thousand kilometres apart. They cannot be undertaken within Europe alone, but the Effelsberg team has already been approached by workers at the Greenbank Observatory in America with this aim in mind.

West Germany's new-found 8000 square metres of precise radio collecting area should also enable her to multiply by at least five the number of known sources in the radio sky to a total of perhaps around 100 000.

8

Parkes telescope looks Down Under

"TRENDS AND DISCOVERIES"
23 November 1961

While British and Dutch radio astronomers pioneered their science in the northern hemisphere in the 1950s, the Australians rivalled them in observing the southern skies never visible from Europe. It was in fact a team of Australian radio astronomers who identified the first radio sources in 1949, and soon after the completion of the big dish at Jodrell Bank, they had a radio telescope almost as large.

The new Australian 210-foot (64-metre) radio telescope was commissioned on 31 October 1961. E. G. Bowen, Chief of the Division of Radiophysics, Australian Commonwealth Scientific and Industrial Organization, is the main instigator of the telescope. It will be part of the new Australian National Radio Astronomy Observatory at Parkes, 300 kilometres west of Sydney – a site chosen for low background radio "noise", flatness and a low average wind speed. The construction of the instrument began in September 1959 and it is now complete apart from the "master equatorial" control system, which will be in operation soon.

The radio telescope – the most powerful in the southern hemisphere – has a parabolic reflector with a focal length of 26 metres. Its surface, accurate to within 9 millimetres, is plated with 10-millimetre steel over a centre section of 16 metres diameter, the rest being covered with wire mesh. The dish is mounted on two short horizontal axes permitting movement in altitude down to 30 degrees above the horizon. The whole rotates as a turret on top of the supporting tower on four tapered rollers running on a hard steel track. It can turn through 450 degrees in azimuth, being driven by servo motors at a rate of 24 degrees per minute. In elevation it moves as 15 degrees per minute. The pointing accuracy is 1 arcminute, which it will maintain in winds of up to 30 kilometres per hour.

Australia has developed a tradition of radio astronomy research. This is its biggest dish, the 64-metre radio telescope at Parkes, New South Wales

Among the variety of phenomena that the telescope will explore are our own Galaxy, the Magellanic Clouds, Jupiter and new radio sources in the outermost parts of the Universe. It should begin to operate at the end of the year.

9

Australians cross the resolution barrier

BERNARD MILLS

3 September 1964

Single big radio dishes are not the answer to all the problems of radio astronomers, and Australian astronomers pioneered a new cross-shaped aerial which "saw" much finer details. This telescope later provided some of the most important results on pulsars, discovered three years after it opened. At the time the telescope was completed, the scientific problems involved were among the least important of the astronomers' worries.

Nearly four years ago it was announced that a Radio Astronomy Centre had been formed in the University of Sydney, Australia, and that the Schools of Physics and Engineering were cooperating to design and build a very large cross-type radio telescope. The instrument is now approaching completion at the recently named Molonglo Radio Observatory, some 30 kilometres distant from Canberra.

A cross-type radio telescope is designed to take advantage of one of the fundamental differences between radio and optical astronomy, the relative importance of sensitivity and resolution (or definition) in obtaining information about the Universe. If one uses a large conventional reflecting telescope, for instance like the Jodrell Bank instrument, for observations at moderate to low frequencies, it turns out that the amount of information which can be collected is limited by the resolution of the radio telescope. If a telescope of the same collecting area is built in the form of a cross, and the arms connected together appropriately, a very much higher resolution is possible and therefore a much greater amount of information can be utilised.

Because of the elongated shapes of the arms, the response patterns of the arms are also elongated. The outputs from each arm are amplified and multiplied together so that only signals picked up by both arms simultaneously appear in the final output. The solid angle over which this correlated response extends is determined by the length of the arms and is much smaller than the total response angle which is determined by the area of the arms. Thus it is possible to tailor the response of a cross-type of radio telescope to suit the particular kind of observation required.

The Sydney University telescope is designed to have very high sensitivity and resolution in the middle region of the radio spectrum while, at the same time, retaining a high degree of flexibility of operation. The arms of the cross are approximately 1½ kilometres in overall length and 12 metres wide; they consist of cylindrical parabolas of wire mesh supported by tubular steel frameworks, and carry two systems of receiving dipoles stretching along each reflector at the line focus. The dipole systems are turned to wavelengths of 73 centimetres and 2.7 metres respectively and may be operated simultaneously. The arms lie accurately in the east–west and north–south directions.

It is a "transit" instrument and each arm is directed independently to the elevation or declination to be observed, the E–W arm by mechanically tilting it about its long axis, the N–S arm by controlling the electrical path length from the 5000-odd dipoles to the central mixing point. Rotation of the Earth then scans the telescope across the region of sky selected. The effective response pattern has an angular diameter of 3 arcminutes at 73 centimetres wavelength and 10 arcminutes at 2.7 metres wavelength. A number of these responses may be produced independently and simultaneously, pointing in different directions so that, like an optical telescope, a "picture" of the part of the sky selected will be produced. Initially this picture will be only ¼ degree across, but it is hoped to increase the size later; no technical difficulty is involved, only finance.

The instrument represents a direct development from the original cross which I constructed some 10 years ago while working with the Commonwealth Scientific and Industrial Research Organization. The performance and general flexibility has been improved by about two orders of magnitude to meet the requirements of present-day research, but, as far as possible, previously established principles and techniques have been followed to ensure the minimum of uncertainty in its performance. However, a major change in technique follows a complete switch from valves to transistors which was necessary because of the much greater complexity of the new instrument.

In arriving at the design, the main astronomical requirements borne in mind have been the need for high-resolution studies of the very rich southern Milky Way fields, as well as the nearest external galaxies, the Magellanic Clouds, which are only visible to southern observers. One of the advantages of the southern hemisphere for astronomy is that we are able to study the latter systems very much more easily than our colleagues in the northern hemisphere can observe the closest northern galaxy, the Andromeda Galaxy, which is some 10 times farther away.

Against these natural advantages the Australian radio astronomer has to contend with a relatively unsophisticated engineering technology and a general lack of interest in basic research among those controlling the official purse strings. These combine to make the financing and construction of a major instrument very difficult. For example, even the government-sponsored 64-metre reflector at Parkes was very substantially financed by American sources, designed by an English firm of consulting engineers and constructed by a German firm.

The present instrument was made possible by an initial generous gift of £80 000 from the Nuclear Research Foundation, a privately endowed foundation within the University of Sydney. Though sufficient to begin the planning and preliminary development work, it was not until the National Science Foundation of the USA made available a grant of $746 000 early in 1962 that a real start could be made.

10

A sharper view from Cambridge

MARTIN RYLE

8 July 1965

Sir Martin Ryle conceived the idea of using the Earth's rotation to "synthesise" a large radio telescope from smaller dishes. After tests, his team constructed the One-Mile Telescope described here. It was effectively a radio dish 1½ kilometres across, but consisting of only three dishes each 18 metres in diameter. The tremendous successes of this telescope in "seeing" details in radio sources led directly to the Dutch Westerbork array, Cambridge's Five-Kilometre Telescope and the American Very Large Array, the "radio eyes" of the 1970s and early 1980s.

Some of the most important questions in current astrophysics relate to the physical processes occurring within radio galaxies and quasars and the source of the fantastic energies involved. It is generally believed that the emission mechanism involves the acceleration of high-energy electrons by magnetic fields within the sources, but the origin of neither the particles nor the enhanced field strength is known. The development of double and more complex sources is not understood, nor is the relationship between radio galaxies and the more compact quasars, if any.

The obvious point of attack on these problems is a detailed study at different frequencies of the distribution of intensity and polarization (which gives information on the magnetic field) within a large number of sources. Such a study has not been possible until now because of the lack of resolution (or sharpness) and of sensitivity in the radio telescopes available. These difficulties have also limited the investigation of cosmology by observations of radio sources.

The new telescope at the Mullard Radio Astronomy Observatory, Cambridge, was designed with these aims in view. It has been constructed with the aid of a grant from the Department of Scientific and Industrial Research (now the Science Research Council). It makes use of the "aperture synthesis" technique, developed at Cambridge over the past 10 years, in which small aerials are moved about to occupy the positions of all the elements of a much larger imaginary one. When the observations are summed by an electronic computer a map of the sky is produced having a resolution equivalent to that of the larger aerial.

Unlike the earlier Cambridge instruments, in which a long array is used in conjunction with a smaller movable aerial, the new telescope employs three 18-metre paraboloidal reflectors, or dishes. These are on an east–west line, two being fixed 760 metres apart and one movable along a 760-metre railway track. They are so mounted that they can be directed towards a chosen part of the sky for a 12-hour period. During this time the rotation of the Earth carries the aerials round each other so that, in effect, two elliptical rings of the larger aperture are built up. By altering the position of the movable aerial, other rings may be added on successive days to synthesise a complete aperture whose diameter is equal to the maximum separation of the elements – 1520 metres in this case. At the present operating wavelengths of 74 and 21 centimetres, the beam widths, which determine the resolution, are then 80 and 23 arcseconds respectively so that, for the first time, it becomes possible to provide a radio map of the sky with a resolution better than the unaided human eye (about 1 arcminute).

In addition to this high resolution, the sensitivity of the "synthesised" aerial is much greater than might be expected from the small size of the elements used. In the present instrument, the sensitivity is equivalent to that of an aerial of diameter 300 metres and it is a surprising and remarkable fact that, using the synthesis technique, no more time need be taken to survey a given area of sky than would be required with the complete large aerial, using the same sampling time.

During the past eight months the instrument has been used for: (1) the determination of the positions of about 50 sources to within a few arcseconds (this is important in order to increase the number of identifications with optical objects); (2) measurements of the distribution of radio brightness across a number of intense sources; (3) a full survey of a sample area, which has revealed sources an order of magnitude fainter than have been observed previously. As with the largest optical telescopes, it is no longer feasible to undertake a survey of the whole sky with the new telescope for this would need more than a thousand years of continuous observation.

The natural successor to the Cambridge One-Mile Telescope is the Five-Kilometre Telescope shown here. Completed in 1972, its four movable and four fixed 13-metre dishes cover a baseline of 4.6 kilometres

In the case of Cygnus A, the source is confirmed as a double object like 80 per cent of the other radio galaxies. Cassiopeia A, which coincides with the faint filamentary remnants of a supernova or exploded star, is clearly resolved for the first time as a ring of emission containing localised peaks and gaps. The simplest interpretation is that most of the radio emission comes from a thick spherical shell. This fits in well with a theory, put forward two years ago by van der Laan, which suggests that the explosion of the supernova compresses the surrounding interstellar magnetic field and traps high-energy electrons which then create radio emissions by the so-called synchrotron mechanism.

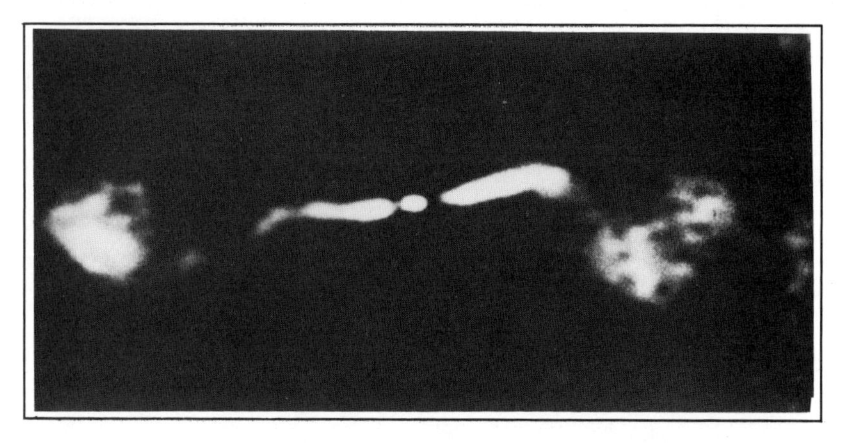

Radio photograph from the Cambridge Five-Kilometre Telescope shows jets of electrons extending a million light years from the centre of the radio galaxy 3C 449 – some of the earliest evidence that the outer lobes of radio galaxies are continuously supplied with energy from a "powerhouse" within the central galaxy

The performance of the new telescope has come fully up to expectations and it is clear from these early results that it will enable the group at Cambridge to continue to make a substantial contribution to some of the outstanding problems in the subject.

11

Merlin conjures up finer resolution

NIGEL HENBEST
18 November 1982

A new interlinked system of radio telescopes centred on Jodrell Bank typifies the radio astronomy of today. The Merlin array uses half a dozen small telescopes to synthesise a "dish" half the size of Wales. It "sees" details one-tenth the size of the finest image in an optical photograph of the sky.

During its 25 years of existence, the famous radio telescope at Jodrell Bank has become to many people the symbol of radio astronomy. Now called the Mark IA, the large dish is still patiently returning important scientific results as it enters its second quarter-century. But this striking landmark has now been joined by a much larger, more powerful radio telescope called Merlin. Unlike the Mark IA, Merlin is not well known because, in line with its name, it is almost completely invisible. While the 76-metre diameter dish of the Mark IA was for many years the largest in the world, Merlin is 133 *kilometres* across. Its invisible "dish" covers an enormous tract of the western Midlands and east Wales (Figure 1).

Merlin consists of six individual radio telescopes, linked electronically so that it mimics a single dish with a diameter equal to the distance between the most widely spaced pair of telescopes: from Tabley just southwest of Manchester (near the M6 Knutsford services) to Defford, near Pershore in Worcestershire. The telescopes all "look" at the same radio source, and send their results by a radio link back to Jodrell Bank, where the data are processed and combined by computer. Each pair of radio telescopes forms an interferometer, and the six telescopes give a total of 15 different interferometer pairs. The whole array is thus a multi-telescope radio-linked interferometer, or MTRLI, as it was first known. But this acronym is difficult to remember, let alone say, so the telescope

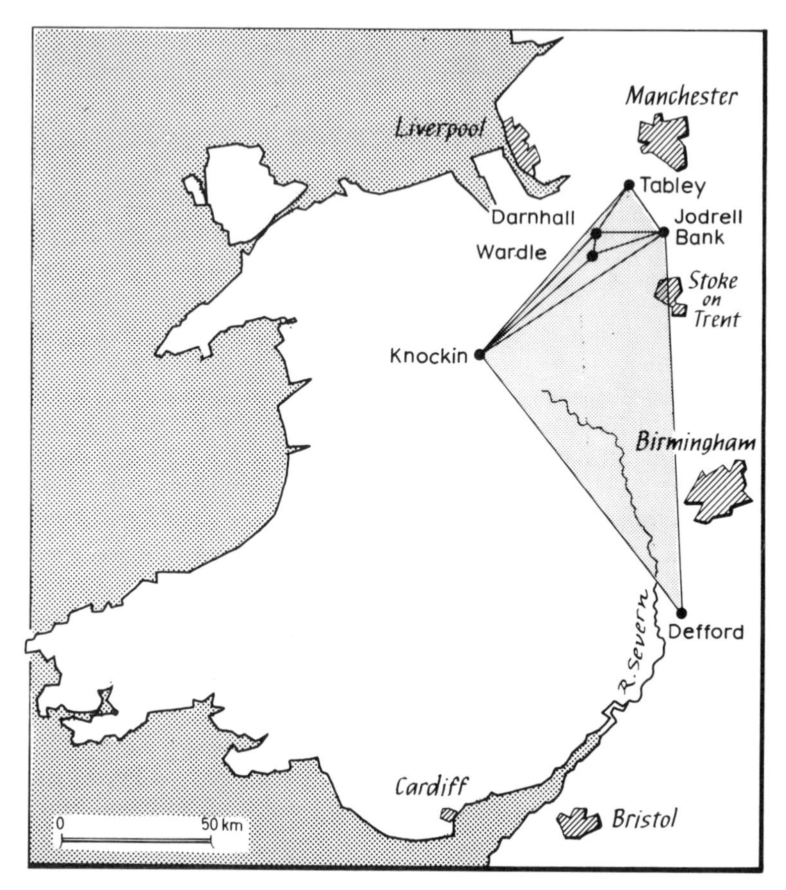

Figure 1 *The radio-linked dishes of the Merlin array of radio telescopes cover much of the West Midlands, and simulate a single telescope 133 km across*

has become a multi-element radio-linked interferometer network, or hey presto, MERLIN!

Radio astronomers obviously have good reason to move from something as simple, if costly, as a single big radio dish to the complexity of Merlin. The driving urge is to greater resolution, the ability to "see" the finest possible details in a radio source. The resolution of any telescope depends on the ratio of the telescope's diameter to the wavelength of the radiation it picks up. If radio astronomers hope to see any detail at all when they observe wavelengths of a metre or so, they must use radio telescopes that are kilometres, or even hundreds of kilometres, across.

A pair of small radio telescopes linked electronically to form an interferometer can show whether a radio source contains fine-scale structure or not. For short separations, or "baselines", the telescopes can be linked by cables; but for longer baselines, it is cheaper and better to use a radio link.

Bernard Lovell encouraged the development of interferometers in the early days of Jodrell Bank, and astronomers there built their first radio-linked interferometer in 1951, over a baseline of 4 kilometres – from Jodrell to Lovell's back garden! They gradually moved to longer and longer baselines, while the Jodrell Bank site itself sprouted the giant Mark IA telescope in 1957 and a few years later the smaller, ellipse-shaped Mark II. By 1967, either of these telescopes could be radio-linked to Jodrell's Mark III, 24 kilometres away at Wardle, near Nantwich, or over 127 kilometres to the radio telescope at Defford. This was one of a pair of radio telescopes that the Royal Radar Establishment had originally constructed at Defford as a short-baseline interferometer to measure positions of radio sources accurately.

These interferometers could register structure in radio sources that was very much smaller than the sizes traditional optical astronomers were used to. The largest optical telescopes should theoretically "see" details much finer than an arcsecond in size, but in practice the Earth's constantly shifting atmosphere blurs out all images to a size of 1 arcsecond or larger. When Jodrell Bank's radio interferometers work at their shortest wavelength, 6 centimetres, the Jodrell–Wardle pair picks up structure on a scale of only ½ arcsecond, while the Jodrell–Defford system responds to details a mere 0.1 arcsecond in extent – the size of this full stop . seen a kilometre away.

This was useful work, and it involved overcoming tremendous technical problems to get the long radio links working correctly. But the two interferometers at Jodrell Bank could not make a map of a radio source. Mapping requires a large number of interferometers, of different lengths, so that each is sensitive to structures of rather different angular sizes; the longest baselines recording the smallest structures, and the shortest telescope separations the broad-scale structure.

The rejection in 1974 of a plan to build a huge new dish in Wales diverted the direction of research at Jodrell Bank away from its emphasis on results from large single telescopes. Rather than strike out in a totally new field, Henry Palmer, a pioneer of interferometers, suggested that Jodrell should stick to the area in which it had built up such technological competence: radio-linked long-baseline interferometry. By adding more telescopes to generate even more interferometer pairs, Jodrell Bank could have enough different baselines to produce real maps. If they worked at the same wavelength as the Cambridge One-Mile and Five-Kilometre Telescopes, Jodrell's longer baselines would show finer details in the maps, while the Jodrell interferometers could play a complementary role by observing at longer wavelengths, and produce the first maps to show details of around 1 arcsecond at wavelengths as long as 73 centimetres.

Under Bryan Anderson, J. G. Davies and Ian Morison, Merlin gradually came into existence. First the astronomers needed a new dish to fill in baselines between those already existing: a telescope at the apex of a triangle would give useful baselines to both Jodrell Bank and Defford. This apex lay near Oswestry, and the dish was erected at the nearby village of Knockin on the Welsh border. With four telescopes, the researchers could test a system that removes one of the worst bugbears of long-baseline interferometry. The signals coming in from space are affected by the atmosphere above a radio telescope, and to a different, constantly changing extent at each site. But it is possible to calculate what is going on, and correct for it, if there are more baseline pairs than there are individual telescopes, using a technique called "closure phase". The number of pairs formed by N telescopes is $N(N-1)/2$; "closure phase" can be brought into operation when there are four telescopes (six pairs) or more.

The success of this four-telescope system gave the green light for two more dishes, sited relatively near to Jodrell Bank at Tabley and at Darnhall, only 6 kilometres from Wardle. These filled in the shortest baselines, essential for showing the large-scale features on the maps; while the Tabley–Defford link became, by a little, the longest in the whole network. The three latest telescopes are identical 25-metre diameter dishes, bought relatively cheaply (about £1 million each) from the United States, where this "E-system" type was being built by the dozen for the latest American radio-telescope set-up, the Very Large Array.

The five outstations are operated by remote control from Jodrell Bank, where each has a control board identical to those that operate the telescopes immediately at hand. A computer can link in either the Mark IA or the Mark II as the Jodrell telescope in Merlin (each is used roughly half the time) or use any combination of telescopes if the entire Merlin network is not being utilised. Each outstation has

In an isolated field at Knockin, Shropshire, stands one of the 25-metre dishes of the Merlin array. As well as returning data by radio link to Jodrell Bank (from the tower in the background), the Knockin dish is remote-controlled from Jodrell

its own computer, which is kept in touch with Jodrell's main computer by telephone land line, while other information, including the telescope's output signal, travels by radio link. Each site is covered by four or five television cameras, partly to show the controller at Jodrell Bank any faults that may arise in the telescope, and partly for security.

One major problem with the radio link is that water vapour on the way affects the atmosphere's refractive index, and so changes the effective length of the radio waves' path. As far as the interferometer is concerned it is as if the distant telescope is moving about erratically. To combat this, the Jodrell Bank computer sends pulses of radio waves to each outstation to be returned immediately: the time taken reveals the effective path length at that moment, and the computer puts a correction into the telescope's output. The changing water vapour along the radio link to Defford and Knockin to the far south and west, coupled with the signals from automatic rain gauges at each site, give astronomers at Jodrell Bank their own unique weather forecast and rain warning. More seriously, the actual length of the well-studied baseline from the telescopes at Jodrell to the dish at Defford can now be calculated to a precision of 1 centimetre in 127 kilometres, and the Ordnance Survey and satellite navigation experts are currently checking their own precision against this exact ruler.

The signals from each interferometer pair are processed at Jodrell Bank, and then sent down to the more powerful computers at the University of Manchester Regional Computer Centre, which convert the interferometer outputs into a map of the radio source. The fact that the telescopes are not in the same line means that Merlin acts as a slightly distorting mirror. Each bright spot on the map produces fainter ghost images, not only positive but also negative in intensity! A special computer program can, however, "clean up" the map, working from the fact that the negative images ("holes in the sky") cannot be real, and can thereby eliminate both the negative and the positive ghost images.

The new results from Merlin have already revealed much about radio galaxies and quasars, the most distant and most powerful radio emitters in the Universe. Earlier observations, particularly at Cambridge and with the similar radio telescopes at Westerbork in Holland, had shown that a radio galaxy or quasar typically has two large radio-emitting clouds of magnetic field and electrons, one on either side of the galaxy itself. They had also hinted that these clouds are fed by diametrically opposed beams of fast electrons, which are ejected from the galaxy's centre and "light up" when they hit the tenuous gas surrounding the galaxy. Astronomers explain these electron beams as coming from either side of a swirling accretion disc of gases orbiting a very heavy black hole, as massive as a thousand million suns, in the centre of the galaxy or quasar.

Merlin has been able to "see" directly the beams in many radio galaxies and quasars, appearing as straight jets sticking out from the centre. But oddly enough, each galaxy or quasar usually has only one jet. The active centre must feed the two clouds not with two simultaneous beams, but by sending out a beam at first in one direction, and then in the opposite direction. It looks as though the black hole must oscillate "up and down" in the accretion disc surrounding it, so that the beam bores its way out only from the side where there is less gas in the way.

This discovery of beams is due to Merlin's high dynamic range: at 18-centimetre wavelength, it can discern faint objects like jets without being dazzled by adjacent quasar cores or "hot spots" that are a thousand times brighter; this is equivalent to the ability to see a bare torch bulb next to an airport runway light. Merlin's uniquely high resolution at long wavelengths also enables it to probe a radio galaxy's history. When fresh electrons are injected into a cloud, they emit radio waves of all wavelengths. But after a few million years the electrons emitting the short radio wavelengths have lost their energy, and so the older clouds are visible only to radio telescopes observing at the longer wavelengths.

Merlin has found that the radio clouds either side of a source are often almost identical, such that any "feature" on one side is matched by a similar feature exactly opposite, on a line through the galaxy's centre. The interpretation is that, in flipping over, the beam has painted exactly the same picture on the intergalactic gases on either side of the galaxy, and that it has been swinging around gradually, to spread out, point by point, identical patterns on either side.

America's new telescope network, the cable-linked Very Large Array (VLA), has also found beams in radio galaxies, but it tends to work at short wavelengths where the "history" is less obvious. Even at its shortest wavelength of 1.3 centimetres, the 27-kilometre diameter VLA falls short of Merlin's resolving power. And although the VLA's 27 telescopes make it more sensitive than Merlin, at present the VLA cannot achieve the same dynamic range. The VLA can detect fainter individual sources, but cannot make out faint details within bright sources. So the telescopes complement one another well. Merlin, however, cost only about one-tenth as much as the $80-million (£50-million) American array.

Merlin has also made discoveries about radio sources closer at hand, within our own Galaxy. It has investigated the radiation from natural masers, the radio equivalent of lasers, which emit radio waves at a wavelength of 18 centimetres from hydroxyl (OH) molecules. The precise wavelength is altered slightly by the doppler effect when the gas is moving, and Merlin can tune in precisely to investigate gas moving at different speeds; it can make high-resolution maps of gas travelling at 80 different velocities simultaneously. Astronomers knew that maser-producing gases surround some old,

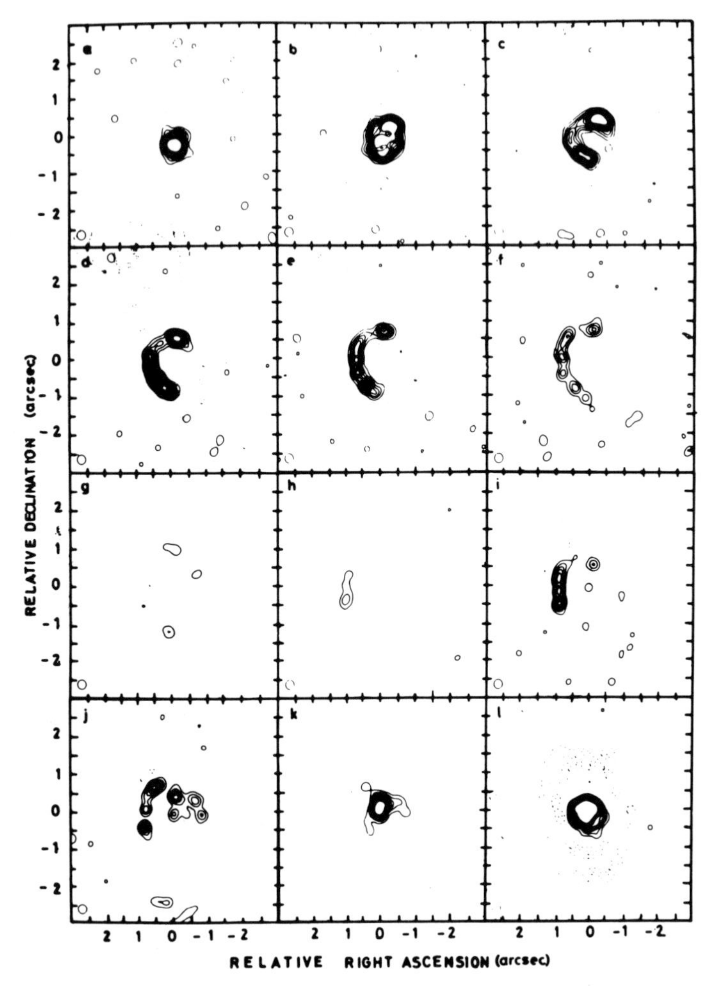

Figure 2 *Merlin shows considerably finer detail than an optical telescope can achieve, as seen in these 12 maps of the same star, made at slightly different wavelengths near 18 centimetres. They show an expanding shell of gases around the star, by the radiation from hydroxyl molecules acting as natural masers. The shell's expansion causes slight doppler shifts, so the maps reveal different cuts through the spherical shell. The first seven maps are at the longest wavelengths, showing the receding back of the shell, successive cuts being slightly nearer to us. The last five are at the shorter wavelengths, revealing the shell's front. The masers amplify the central star's radiation outward and inward, so we can detect the back and front of the shell, but not the sides*

red-giant stars. Merlin has found that this gas is arranged in a way that is almost too simple for hard-bitten astronomers to believe. It forms a simple, spherical expanding shell about 10 times larger than the solar system and, as the simplest models predicted, it appears brighter at the centre than at the edge (Figure 2).

With Merlin's phenomenal success in its first couple of years, astronomers at Jodrell Bank are keen to add even longer baselines, to produce maps showing more detail. A 200-kilometre link to the dishes of the One-Mile Telescope at Cambridge would give Merlin a better east–west coverage, and improve its resolution at 6-centimetre wavelength to $^1/_{20}$ arcsecond.

Merlin's resolution could also be improved by working at shorter wavelengths, but only the three new E-system dishes at Knockin, Darnhall and Tabley are constructed accurately enough for this. Current plans are to link these together with another dish now being installed at Jodrell Bank, a 13-metre dish which has come from a radar installation at the Woomera rocket range in Australia. These four telescopes could investigate the natural water (H_2O) masers, radiating at 1.3 centimetres, in finer detail than the whole Merlin array studying the longer wavelengths of the hydroxyl maser at 18 centimetres. Ideally, Merlin needs more modern dishes to replace the older telescopes, to fill in more intermediate baselines (with perhaps a dish near Shrewsbury) and to extend the array.

But Jodrell's future does not lie only with Merlin. The old Mark IA is a very powerful telescope for many purposes, and is still the world's second-largest fully steerable radio dish. Its work includes measuring the speed of pulsars in our Galaxy, and "weighing up" distant galaxies by their motions. Merlin also uses the big dish for roughly half the time, when observing very faint sources. The Mark IA is also a cornerstone of the European Very Long Baseline Interferometry (VLBI) network. In this much larger version of Merlin, signals are recorded at each telescope independently, and the magnetic tapes are taken to Bonn to be combined later. Europe's VLBI network is much more sensitive than that operating in the United States, because it utilises the 76-metre Mark IA and the 100-metre Effelsberg telescope, along with the large collecting area of the Westerbork array. The workload of short-baseline arrays like that at Westerbork is now shifting to the large networks like Merlin and the VLA, and the 14 dishes at Westerbork are regularly used as the equivalent of a 93-metre single dish for VLBI work.

The European network will become increasingly important in years to come, with new "Mark III" receivers and analysers, and two purpose-built Italian radio telescopes at Bologna and in Sicily. The network can also use, part time, radio telescopes in Sweden, the Crimea, South Africa and, possibly in the future, in Spain and the east coast of the United States. The Mark IA will then be a fundamental part of a radio telescope as large as the Earth. Radio astronomers are now discussing with the American space agency NASA the possibility of a 15-metre radio telescope in orbit about the Earth, which would extend the baselines three or four times more. This network would be able to make out the details of the swirling gas accretion discs in the nearest radio galaxies.

It is difficult to predict more than 10 years ahead in the rapidly changing world of astronomy, and Graham Smith – Jodrell's present director, and Astronomer Royal – would not hazard a guess at what would be happening at Jodrell Bank more than a decade from now. But looking through the window of his study at the great white face of the Mark IA, he remarked affectionately, "That darn thing's been going 25 years, and I reckon it's got another 25 years of useful work ahead."

12

Turning continents into radio telescopes

"THIS WEEK"

25 March 1982 and 30 October 1980

In late 1982 the Australian government approved the construction of a "Merlin Down Under", a 300-kilometre array of radio telescopes to be completed by Australia's bicentenary in 1988. It could later expand to cover the entire continent. Meanwhile, two other proposals are afoot to turn North America into a huge radio telescope.

Despite Australia's early lead in radio astronomy, it did not progress to building arrays like Britain's Five-Kilometre Telescope and Merlin, or the American Very Large Array. In recent years Australian astronomers have been pushing for a 6-kilometre array, consisting of the existing big dish at Parkes and five new dishes in line with it. This "Australian Synthesis Telescope" would have cost £10 million; it was turned down in the 1981 science budget.

Now astronomers Down Under have new plans for a larger "Australia Telescope", which would cost £15 million. This telescope would have three elements. One would be a purpose-built array of five 22-metre dishes at Culgoora, near Narrabri in New South Wales. An existing radio telescope at Culgoora is scheduled to close in 1984, so the new array could benefit from existing support facilities there. Another 22-metre dish would be built at the Anglo-Australian Observatory at Siding Spring, the site of Australian and British optical telescopes. The third element would be the 64-metre dish at Parkes.

The Culgoora array on its own could serve as a 6-kilometre telescope to map the broader features of sources. Linked to the other two dishes, it would form an array equivalent to a single dish 300 kilometres across. Such an array would be able to make useful observations of the astronomical objects that have been discovered by the new generation of satellites carrying telescopes and other observing instruments that operate at other wavelengths.

The Australia Telescope could be complete by 1988, the year Australians celebrate the bicentenary of the first settlement of their continent.

But there are also ideas afoot even more ambitious than plans for a 300-kilometre array. The Australia Telescope could be linked by satellite to other radio telescopes at Tidbinbilla (near Canberra), Sydney, Alice Springs, Hobart in Tasmania and Carnarvon in Western Australia, to create a huge array spanning the whole continent – a radio telescope 3000 kilometres across. This array could resolve details a thousand times smaller than the features that even the best optical telescopes can see. It would also nicely complement the transcontinental arrays that Canadian and American astronomers are independently planning in North America.

A team from Caltech and NASA's Jet Propulsion Laboratory in Pasadena has completed a feasibility study of a network covering the United States. The study envisages 10 identical radio "dishes", each 25 metres across. They would all observe the same radio source simultaneously, and a central computer would combine their results to produce a map showing details as fine as would be revealed by a single telescope as wide as the American continent.

The new instrument should enable astronomers to discover just what is happening inside quasars – the brightest objects in the Universe, where the power of a million million suns is packed into a volume no larger than the solar system.

Radio astronomers already use such techniques – called very long-baseline interferometry – between existing radio telescopes in the United States. But the present link-up is far from ideal. The dishes are different; they produce different outputs and they are scattered irregularly over the country. And there are administrative difficulties.

The new, purpose-built network would suffer none of these problems because it would operate as a single installation. The dishes in the network would be similar to those installed at the Very Large Array (27 kilometres across) in the New Mexico desert near Socorro. But they would be cheaper because – unlike the dishes in the VLA – they would not be mobile. The study recommends sites spread widely across the United States from Hawaii to Massachusetts and from Alaska to Texas.

The feasibility study estimates the cost of the network would be

£22 million – about half that of the VLA – and running costs would be £3 million per year.

Meanwhile, the Canadian Astronomical Society has its own plans for a coast-to-coast network of eight radio telescopes. The network would extend along latitude 49.3°, from Gander, Newfoundland on the Atlantic, to Courtenay, British Columbia on the Pacific, a distance of 4200 kilometres. The society proposes that the middle six antennas should be in Cochrane, Ontario; Weyburn, Swift Current and Medicine Hat, in Saskatchewan; and two in Lethbridge, Alberta. Different combinations of these antennas would give a variety of baselines, creating the flexibility to observe both large and small radio sources. If necessary, two existing Canadian radio telescopes, at Algonquin Park, Ontario, and Penticton, British Columbia, could be linked into the system. Individual telescopes would again be similar to those in the Very Large Array in New Mexico.

PART FOUR

The View from Space: Satellites for Short Wavelengths

One minute before midnight on 18 June 1962, an Aerobee rocket blasted off into the warm air of a New Mexico night, carrying a trio of Geiger counters. During its short flight it detected X-rays coming from a far-off source – later called Scorpius X-1. This flight opened up a third window on the Universe. Thirty years earlier, the radio pioneer Karl Jansky had detected cosmic radiations millions of times longer than light; now astronomers could study the Universe at wavelengths a thousand times shorter than the optical view.

The long delay between the beginning of radio astronomy and that of X-ray astronomy was partly due to the fact that astronomers generally did not expect to pick up any radiation at the short wavelengths, at least with the detectors available a few decades ago. The Sun's X-radiation had been detected in 1949, but that was only because the Sun is so close to us on the cosmic scale. Other stars with the same intrinsic power would have been much too weak to have registered on the early Geiger counter detectors. It was largely because radio astronomers had revealed unexpectedly powerful sites of astronomical activity, in supernova remnants and radio galaxies, that some pioneers were prompted to look for X-rays from space. Their optimism was justified, for sources like Scorpius X-1 radiate thousands of times more power in X-rays than the Sun produces at all wavelengths. The spirit of conservatism was generally still so strong in 1962, however, that the successful rocket flight was funded ostensibly to seek X-rays from the Moon, in order to aid the American Apollo programme.

But there is a major natural barrier to overcome. Our atmosphere absorbs all short-wavelength radiations from space. Although some observations can be made from high-flying balloons, the study of these radiations in detail had to await the development of rockets, and later the greater observing possibilities of satellites in long-lived orbits.

Artist's impression of NASA's Einstein X-ray astronomy satellite. Launched as HEAO-2 in November 1978, the satellite produced the first X-ray "photographs" of astronomical objects and was renamed in honour of Albert Einstein, whose centenary occurred in 1979

The shortest wavelengths of visible light are violet, at about 390 nanometres. Moving to wavelengths too short to be visible to the human eye, we come to the ultraviolet domain. Slightly shorter wavelengths than this can in fact penetrate the atmosphere and be recorded by a normal optical telescope and detectors. To astronomers, the ultraviolet runs shortwards of 310 nanometres, the edge of the atmosphere's "optical window", into the wavelength region where ultraviolet is absorbed by ozone in the stratosphere. For wavelengths shorter than 200 nanometres, atmospheric absorption is due to molecules and atoms of oxygen and nitrogen at higher levels, up to 150 kilometres above the Earth's surface. This absorption extends to X-ray wavelengths, rather arbitrarily defined as radiation of wavelength less than 10 nanometres. Although we think of X-rays as "penetrating" radiation, remember that the weight of atmosphere above our heads is equivalent to that of 760 millimetres of mercury – enough to absorb X-rays very thoroughly!

Shorter wavelength radiations are composed of individual packets, "photons", of successively higher energies. X-ray astronomers often talk in terms of energy rather than wavelength, using the unit of the keV (1000 electron volts). The wavelength in nanometres is roughly the reciprocal of the energy in keV (to be precise, multiply the reciprocal by 1.24). To astronomers, radiation of wavelength less than 0.01 nanometre (energy greater than 100 keV) are gamma rays. At these very short wavelengths and high energies, it makes much more sense to use energy units, in particular the MeV (1000 keV). The highest-energy gamma rays can penetrate our atmosphere down to the stratosphere, and can be observed by balloon. The best results come from satellites, however, because they can operate continuously for long periods and collect a relatively high number of the rare gamma-ray photons from space.

The cosmic sources of ultraviolet and X-rays are generally regions of hot gas. All objects emit *black body* radiation whose maximum intensity comes at a wavelength which is inversely proportional to the body's temperature. Ultraviolet radiation originates in stars and gas clouds at temperatures of between 30 000 K and 1 million K; while X-rays are produced in gas clouds with temperatures between 1 million K and 1000 million K. The latter generally occur either in clusters of galaxies, or where gas is heated up as it spirals inwards in the strong gravitational field of a neutron star or a black hole. Nothing in the Universe, however, is hot enough to produce gamma rays in this way. They come instead from *non-thermal* processes like the collision of cosmic rays with the nuclei of gas atoms in space.

Here we look in particular at three satellites, one operating at each of these wavelengths: IUE, Exosat and COS-B. All three are European, or European-inspired. This reflects the fact that space astronomy is no longer the preserve of the United States, as it was in the 1960s and 1970s. In the 1980s American astronomers are facing budget cutbacks, with much of their available money committed to the (mainly optical) Space Telescope. Other countries are now planning major projects in space astronomy that will not be overshadowed by American efforts. As well as the projects of the 11-nation European Space Agency, satellites are being built by Germany and Japan, while France is collaborating with the USSR in orbiting telescopes which match Soviet rocket power with western electronic sophistication.

The International Ultraviolet Explorer (IUE), developed and operated by the USA, UK and ESA, pioneered the way to international use of a space observatory. Its spectral studies should continue to the mid-1980s. There will be some follow-up from the Space Telescope, due for launch in 1986, as this will include some ultraviolet spectroscopy among its many activities at ultraviolet as well as optical wavelengths.

Exosat, the latest X-ray satellite, is the successor to the epoch-making Einstein Observatory, an American satellite which carried the first telescope to focus X-rays. It could "see" details as fine, and objects as faint, as the world's largest optical telescope. At its launch in 1978, the Einstein Observatory marked an advance in X-ray astronomy in only 16 years that paralleled the development of optical

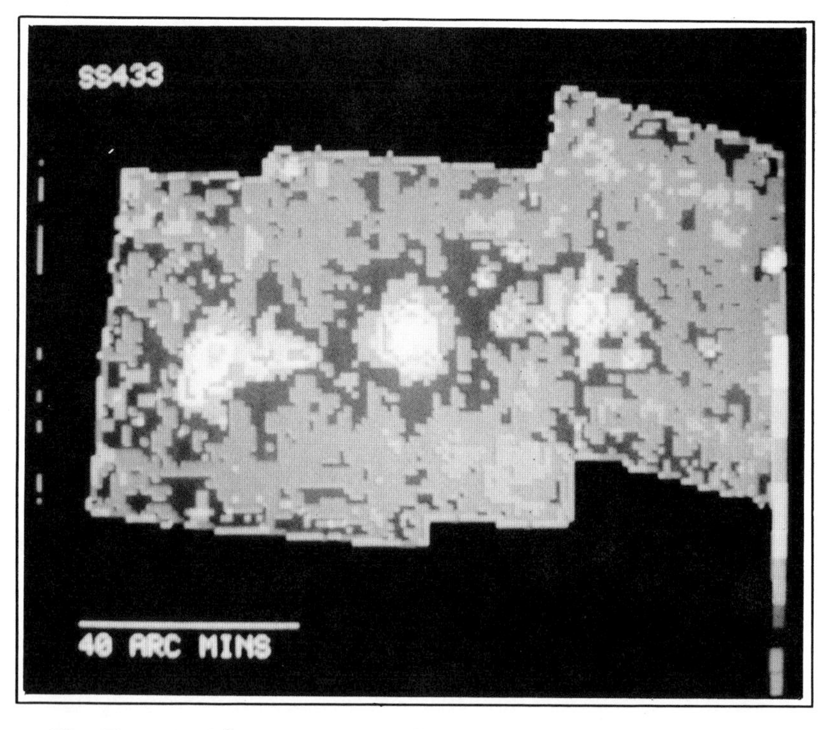

The Einstein Observatory revealed jets of hot gas extending from SS 433, one of the strangest stars in our Galaxy. SS 433, at the centre here, probably contains the core of a star which exploded 40 000 years ago. Now a neutron star or a black hole, this core is pulling gases from a companion star – and ejecting some of the gas as two jets travelling at one-quarter of the speed of light

astronomy from Galileo's telescope of 1609 to the opening of the Palomar 5-metre (200-inch) telescope 340 years later!

X-ray astronomy has a very active future. Exosat will be followed in 1987 by the larger X-ray telescope on the German Rosat (*Röntgen-Satellit*), and in the 1990s by a similar Japanese satellite. British astronomers are designing a large X-ray satellite devoted to spectroscopy, "X-ray astronomy's IUE". The next step forward in imaging X-ray sources should come in the mid-1990s with America's huge Advanced X-ray Astrophysics Facility, a £300-million project still awaiting approval.

Ultraviolet and X-ray astronomers are also tackling one of the still-unexplored regions of the spectrum, the extreme-ultraviolet wavelengths of 10–91 nanometres. Until recently, astronomers thought that this radiation would be totally absorbed by hydrogen atoms in interstellar space, but it now seems that some directions at least are in fact clear. The pioneer instrument to observe sources at these wavelengths will be a British telescope mounted on the German Rosat X-ray satellite.

At the shortest wavelengths, the COS-B gamma-ray satellite will be followed in the late 1980s by the Soviet–French Gamma-1 and the American Gamma Ray Observatory. They will pinpoint and help to identify the two dozen strong gamma-ray sources found by COS-B, most of which – to astronomers' surprise – do not coincide with any obvious optical, radio or X-ray source.

13

The first space telescope

ANDY FABIAN

19 July 1973

The third of the American Orbiting Astronomy Observatories was named Copernicus at the 500th anniversary of the birth of that pioneer astronomer. The satellite carried one of the largest telescopes ever launched, on a space platform whose steadiness has still to be bettered. The Copernicus satellite operated for a record eight years, until 1980, and was closed down only because its simple ultraviolet detectors had looked at every star bright enough for them to see. Copernicus also carried the first X-ray detectors to concentrate the radiation by reflection – a precursor to the true focusing telescopes of the Einstein Observatory and its successors like Exosat.

The Copernicus satellite, launched on 21 August 1972, carries an ultraviolet telescope from the Princeton University Observatory, and X-ray telescopes from University College, London. In less than a year of operation, a wealth of interesting and exciting data has been amassed on these new fields of astronomy. The ultraviolet experiment is proving an effective collector of information about the interstellar medium. Spectrophotometric analysis of the ultraviolet emission from stars enables the absorbing atoms and molecules in that medium to be charted. X-ray maps, source positions and spectral and timing information from the X-ray detectors are opening up our knowledge of the X-ray universe. The British X-ray experiment essentially rides piggy-back, since the main body of this 2-tonne, 5-metre-long satellite is occupied by the 80-centimetre-aperture Princeton ultraviolet telescope.

Due to the rigorous checking and testing of a satellite experiment, which has to operate in space for several years, the basic equipment and experiments were decided upon over 10 years ago. The satellite is one of the Orbiting Astronomical Observatories, being an observatory in the true sense of the word since it can point and remain locked on any chosen region of sky for several hours.

The satellite is pointed at specific regions of the sky primarily by an inertial reference unit. The system is updated on a bright star, the position of which has been measured accurately (to a fraction of an arcsecond) by earthbound telescopes. It then slews off at up to 6 degrees per minute to the required position, by means of coarse and fine inertial wheels. These save the use of the gas jets, which are only needed if the spacecraft loses control. Fifty per cent of the light incident on the Princeton telescope is used for guidance, and enables the system to be sensitive to stars at the limit of naked-eye visibility. Although continual updating of the guidance system is possible when observing a bright star, no known X-ray sources are bright enough in the visible or ultraviolet for the guidance sensors. However, the inertial wheels, which operate in approximately 20-arcsecond steps, allow pointing repeatability to less than 30 arcseconds. When locked on, drift rates are only 2 arcseconds per hour.

The ultraviolet and X-ray experiments are somewhat incompatible in the targets that they observe. Moreover, the ultraviolet spectrometers take several days to complete their scans to the desired accuracy. The compatibility of the experiments lies in the stable viewing platform they both require, which approaches that of large earthbound telescopes. The observing time is divided such that the X-ray observers choose the target for only one day in 10.

The X-ray detection on board the satellite consists of two tiny proportional counters at the foci of grazing incidence telescopes, which operate over the wavelength ranges 0.3–0.8 nanometres and 0.8–2.5 nanometres. These telescopes act as X-ray collectors; they do not image the photons because of the large amount of chromatic aberration produced by the grazing incidence requirement for the reflector surfaces. Nevertheless, adjustable apertures can restrict the fields of view of the detectors to 10, 3 and 1 arcminutes and 10, 6 and 2 arcminutes, respectively.

14

Ultraviolet astronomy comes of age

JON DARIUS
12 April 1979

The International Ultraviolet Explorer (IUE) ranks as one of the most successful and longest-lived satellite observatories, in 1984 still returning important data after six years in orbit. Only a year after launch, IUE had already made outstanding new discoveries.

IUE is not the first satellite to undertake ultraviolet spectroscopy, but it is arguably the most versatile. In the wake of early rocket studies, some dozen or so ultraviolet experiments have been carried aloft by satellites over the past 15 years. Different experiments catered for the specific requirements of various groups of astronomers. But, while no ultraviolet satellite can be all things to all astronomers, IUE does operate over the widest spectral range yet, covering wavelengths from about 115 to 320 nanometres. It also allows the user to choose between high-resolution operation – which gives a detailed spectrum, but is time-consuming – and a low-resolution mode. Indeed, this is an example of IUE's most novel feature; it is the first telescope put into space that can be controlled in the same way as a ground-based telescope, by astronomers visiting its "home" station. There observers can pick out the "target" stars of their choice and direct the activities of the spacecraft.

But aside from any scientific records that IUE may be setting, it assuredly holds the world title for the longest time between initial

---

(Opposite) *The International Ultraviolet Explorer satellite, launched in 1978, is still sending back very high-quality spectra of hot objects in space. The spacecraft is controlled both from the Goddard Space Flight Center, Maryland, and by European astronomers in Madrid*

conception and launch. Conceived as the "Large Astronomical Satellite" in a British design study commissioned by the European Space Research Organisation in 1964, axed on financial grounds in 1967, reborn as the "Ultraviolet Astronomical Satellite" only to be re-axed, the stillborn satellite seemed to imply that the British contribution to ultraviolet astronomy would be restricted to Skylark rockets and balloons. At this point, Robert Wilson (then at Culham Laboratories and the future project director in this country) turned to the American space agency NASA. "How do you sell a rejected European satellite to the foremost space agency?" he muses retrospectively. "You can only give it away."

The ultraviolet phoenix was soon reborn, this time with NASA's Goddard Space Flight Center fanning the flames. European interest was re-awakened, and by the end of 1971 the future of the satellite, finally christened the International Ultraviolet Explorer, was assured. Launch, spacecraft and scientific payload were the responsibility of NASA, except for camera design and construction (the British contribution, financed by the Science Research Council) and solar panels (contributed by the European Space Agency). The three agencies agreed to treat IUE as an international observatory, so that an astronomer awarded observing time was to be responsible for gathering his own data by directing operations at the ground station. This makes it quite unlike previous satellites where a scientist based in a university was wholly isolated from the actual process of collecting the data.

It was not only the 15-year gap between conception and birth, and the usual hopes and fears of ultraviolet astronomers, which occasioned a great deal of pre-launch nail-biting. Delta rockets like the one intended for IUE had chalked up two disastrous failures in the preceding year despite an excellent track record before that. In the event, however, IUE was launched without a hitch by a Delta 2914 from Cape Canaveral on 26 January, 1978. "Lift-off late but perfect", crowed one American headline. The satellite achieved mission orbit on 28 January, acquired its first star on 6 February, and then observed high-priority targets for the next three weeks just in case of a premature shut-down. After final optimisation and calibration of the instrumentation on board the spacecraft, IUE opened as an international observing facility on 3 April 1978.

The IUE satellite stands 4.2 metres high from exhaust to sunshade, and altogether weighs 671 kilogrammes. It operates on solar power provided by two solar paddles of silicon cells or, when short of sunlight, by two nickel-cadmium batteries. The solar array, generally oriented at right angles to incident sunlight, yields 400 watts although the average power requirements run to 210 watts. During eclipse seasons – three-week periods of daily hour-long solar eclipses which occur at half-yearly intervals – the batteries come into play.

The scientific payload of IUE consists basically of a telescope, mechanisms for fine guidance of the satellite, two spectrographs and television cameras. The 450-millimetre aperture telescope feeds the ultraviolet radiation to one of the two spectrographs – instruments which record the spectrum of any ultraviolet source in the telescope's field of view. They operate over two different ranges in wavelength: 115–195 nanometres, and 185–335 nanometres. A camera in each spectrograph converts the ultraviolet image to video signals suitable for transmission to ground. The faceplate of the television tube is scanned on command when an exposure is complete, and the picture is replayed to one of two tracking stations – the Goddard Space Flight Center in Maryland, or the Villafranca Satellite Tracking Station near Madrid. To prepare for the next exposure, the faceplate is uniformly irradiated with an incandescent floodlamp and then scanned to remove all trace of the previous exposure.

The observer can obtain detailed "high-dispersion" spectra in which the light is split according to wavelength with a resolution of about 0.01 nanometre, or he can produce "low-dispersion" spectra with a resolution of 0.6–0.9 nanometre. Hot stars as faint as 11th magnitude are routinely imaged at high resolution, while much weaker sources can be recorded in the low-dispersion mode. The current record is the 17th magnitude quasar designated Q2204–408.

Star tracking is achieved by fine error sensors – cameras which allow the observers to train the telescope on their desired "target" stars. Each produces an *optical* scan of the sky in a 16 arcminute square around the target. Armed with a corresponding sky chart, an astronomer can identify his target, manoeuvre it into the desired camera, and record its spectrum in the knowledge that the six gyroscopes, together with the tracking mechanism, ensure that the telescope is guided with an accuracy better than 1 arcsecond. A circular image of a freshly exposed spectrum appears on the television monitor in a matrix of points – like a picture on newsprint – with 768 × 768 picture elements (pixels). The signal for each pixel is digitised (numerically coded) into one of 256 levels varying from black to white, or colour-coded.

Time-sharing among the three agencies operating the satellite roughly reflects funding. American observers have the lion's share, 16 hours a day, and the remaining 8-hour shift is accorded alternately to British and continental astronomers. The satellite hovers 42 000 kilometres (on average) over the mid-Atlantic in an elliptical orbit designed to keep it above the same spot on the Earth's surface. It is monitored continuously from Goddard, and for over 10 hours each day from Villafranca. At such a relatively lofty altitude, the Earth appears only 17 degrees across, and, save for constraints imposed by the Earth, Moon and Sun, IUE can freely scour the whole sky. Pointing within 43 degrees of the Sun is prohibited by the sun shade, which also contains baffles to protect the telescope from scattered sunlight.

So much for the where, when and how; but why are ultraviolet observations so vital and what results do they reveal which could not be gleaned from the visible region of the electromagnetic spectrum? The most potent argument for ultraviolet astronomy is the remarkable fact that the strongest spectral lines of even the commonest atoms and ions lie primarily in the ultraviolet. The seven most abundant elements in the known Universe are (in decreasing order of abundance) hydrogen, helium, oxygen, carbon, nitrogen, silicon and magnesium – and not one of these displays its strongest lines in the visible region. The most intense lines that astronomers see in the visible spectrum of solar-type stars are due to sodium – the so-called "D" lines around 589 nanometres – and singly ionised calcium (atoms with one electron removed) with lines at 393.4 nanometres and 369.9 nanometres. To discover other elements we must consult the ultraviolet spectra.

Other powerful arguments abound for launching rockets, balloons and satellites above our ultraviolet-opaque atmosphere. A stellar astronomer would point at once to the fact that hot stars radiate the bulk of their energy in the ultraviolet, because the peak in their emitted radiation shifts to shorter wavelengths at higher temperatures. Moreover, spectral lines from the outer atmospheres of stars – the chromospheres and coronae – are only accessible in the middle to extreme ultraviolet.

Cosmologists have a particularly compelling case in support of ultraviolet astronomy on two grounds. First, the best measurements of the cosmic deuterium ("heavy" hydrogen) abundance come from a series of lines that are displaced by a fixed amount from the normal hydrogen lines. The importance of deuterium resides in its probable primordial origin – in other words, it is not a product of stellar nucleosynthesis (the nuclear "burning" process by which elements are formed in stars), but of the big bang itself (along with hydrogen and helium). The quantity of deuterium created at early epochs reflects rather sensitively the density and amount of primeval matter, and, by extrapolation, the overall density now – a number which, according to theoretical cosmological models, will determine whether the Universe is "open" or "closed" – whether it will continue to expand for ever, or eventually collapse into a "big crunch".

Secondly, there is the controversy waged on one hand by upholders of the principle that redshifts serve as distance indicators, and on the other by debunkers who consider them partly or wholly non-cosmological. This argument could be settled by ultraviolet observations. Quasars manifest extremely large red shifts which, when interpreted in terms of recession velocities according to the majority view these days, imply that they are the most distant objects observed in the cosmos. Certain iconoclasts nevertheless point accusingly at apparent associations between objects of very different red shift and thereby discredit red shifts as cosmological yardsticks. I shall later explain the importance of ultraviolet studies in resolving this debate.

In its first year of observations, IUE has returned a wealth of data to its ground stations. Some of its most important results relate to the interstellar gas in our Galaxy. Without access to ultraviolet spectral lines, astronomers could only speculate about the interstellar abundance, distribution and physical parameters of species other than atomic hydrogen. A host of interstellar lines, many recorded for the first time, are now visible "in silhouette" against the spectra of distant stars. They include lines of multiply-ionised carbon, nitrogen, aluminium and silicon, arising in a continually ionised hot gas. IUE has also revealed highly excited interstellar gas in the halo of our Galaxy. The abundances of metals in the galactic halo appear to be one or two orders of magnitude below solar abundances. Perhaps the halo gas still bears the mark of primordial abundances, from an epoch before the relative amounts of each element were influenced, and the metals enhanced, by the stellar nuclear processing such as occurred in the gas in the galactic disc. The interstellar medium in a neighbouring galaxy, the Large Magellanic Cloud, reveals itself as a broad spectral feature centred on 220 nanometres. This phenomenon is already familiar to astronomers from its widespread occurrence in our own Galaxy, and probably arises from fine graphite dust.

IUE has investigated many individual objects. A long exposure of the Crab Nebula reveals the spectrum of a point source – probably the Crab pulsar – as well as the nebula's continuum spectrum. It has caught outbursts of novae (notably Nova Cygni 1978) and has followed their evolution. Simultaneous observations by IUE (in the ultraviolet region) and the Anglo-Australian Telescope (in the visible region) of the providential eruption of the dwarf nova VW Hyi in January 1979 beautifully confirmed the power-law spectrum which had been predicted.

The satellite detects hot stars to a much fainter limit than previously possible. Vital lines of helium, carbon and nitrogen in the 145–180 nanometre region, inaccessible to the Copernicus telescope aboard OAO-3 launched in 1972, can be studied with com-

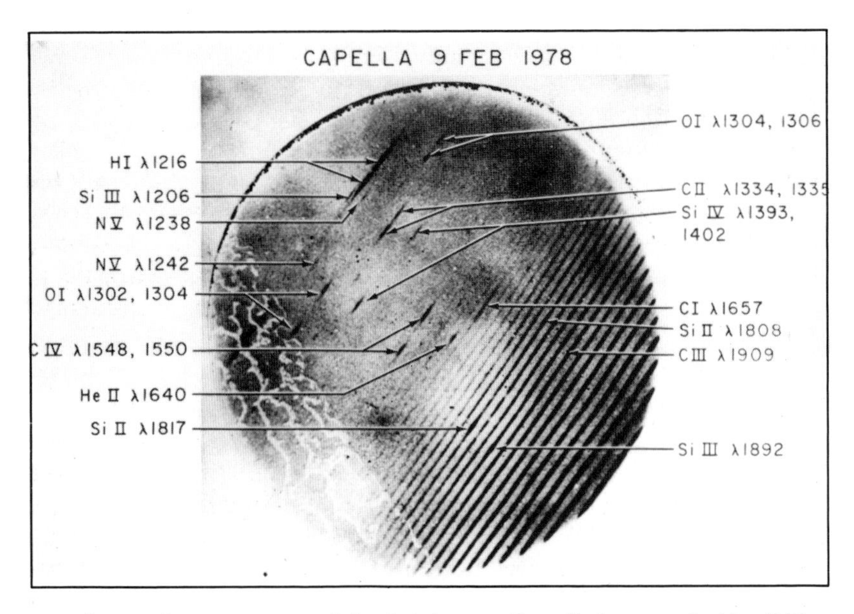

Ultraviolet spectrum of the bright star Capella is recorded by IUE as successive diagonal strips. Each strip covers a short-wavelength range, with wavelengths increasing from top right to bottom left; strips from left to right are at successively longer wavelengths. The strongest spectral lines are marked with the symbol of the element that produces them, with their wavelengths in angstroms (divide by 10 to convert to nanometres). In this negative photograph, the continuous dark strips (lower right) are radiation from the star's surface. The isolated dark patches at upper left are emission lines at shorter wavelengths from Capella's much hotter atmosphere

parably high resolution. The relatively bright hot subdwarf designated BD + 75°325 displays a well-developed series of lines of ionised helium, which are related to a similar series of lines at infrared wavelengths produced in other stars by hydrogen. Although the star BD + 75°325 shows no evidence of mass loss, other subdwarfs reveal an expanding atmosphere like that of their more luminous counterparts. The white dwarf HZ 43, detected in the extreme ultraviolet (most strongly at 30 nanometres) during the Apollo–Soyuz mission, displays a hydrogen line of stellar origin but no other notable spectral features, as we expect for a hot white dwarf where gravitational separation has already taken place with hydrogen left at the surface as heavier elements have sunk inwards.

Cooler stars also yield dividends. The first high-resolution spectrum of Capella provided the most extensive set of data on a stellar chromosphere obtained for any star other than the Sun. And in two variable stars the chromosphere was found to attain a temperature of 250 000 K, two orders of magnitude higher than the cool underlying photosphere.

Simultaneous observations, using IUE, other satellites and ground-based instruments, allow the study of various sources across their whole spectrum, from X-ray to radio wavelengths. The aim is to study variability in stellar winds from supergiants, extragalactic objects which vary violently, such as Markarian 501, and above all binary X-ray sources. The model currently fashionable for binary X-ray sources is built on a system in which mass is transferred from the optical (visible) star onto an extremely dense, unseen object – probably a neutron star or black hole. IUE has discovered a stellar wind streaming at up to 2500 kilometres per second from a star and compact companion together known as HD153919, and producing a circumstellar bubble between the shock-heated outflowing gas and the cold interstellar medium.

Among extragalactic objects studied with IUE are spiral, elliptical and irregular galaxies, Seyfert galaxies (which have small bright nuclei), quasars, and BL Lac-type objects (variable galactic nuclei, similar to quasars, but with no lines in their spectra). A continuum spectrum from the active nucleus of the bright Seyfert NGC 4151 is seen superimposed on its thermal continuum spectrum from stars. Strong emission lines – the so-called "forbidden" lines – provide information on the dust content and on local abundances. The abundances of carbon, nitrogen and oxygen appear to be similar to those in the Sun. In another Seyfert galaxy, NGC 1068, nitrogen appears four to six times too abundant. Two BL Lac objects display a line-free ultraviolet spectrum as at visible wavelengths. The absence of lines seems to represent a deficiency in gaseous material which can be ionised.

Returning to quasars, we paradoxically gained an unimpeded view of their ultraviolet spectra well before "satellite" astronomy got off the ground; their redshifts – many of which when naively interpreted seem to imply recession speeds faster than light – can push the wavelengths of the ultraviolet lines into the visible region. Thus the ultraviolet spectra of "distant" quasars (large redshift) and optical spectra of relatively "nearby" quasars (small redshift) are both visible to us, but to obtain the ultraviolet spectra of the latter we must resort to detectors above the atmosphere. Would the quasars with small redshifts display, in addition to their strong emission lines, the absorption lines observed in the spectra of quasars with larger redshifts? Because the absorption lines have lower red shifts than the quasar emission lines they presumably arise from matter lying between Earth and the quasar.

The brightest quasar, 3C273, does not seem to display absorption lines from intervening galactic or intergalactic matter as observed in the ultraviolet spectra of more distant quasars. The absorption lines we do see can be attributed to matter within our own Galaxy – they have zero redshift. Also the putative dip in the hydrogen continuum spectrum, which could be produced by unresolved absorption lines from intervening objects of lower redshift, is seemingly absent – a striking confirmation of the cosmological hypothesis for quasar red shifts.

The incipient impact of IUE on both observational and theoretical astronomy augurs very well indeed. The amount of data received at the European station alone is quite staggering: in the past year, some 2000 spectral images were taken by 91 observing groups from 11 countries. The spectral total for both stations in the first year of operation was more than 6300 and this figure is confidently predicted to rise in the coming year. Ultraviolet astronomical spectroscopy has never been healthier.

15

An ultraviolet eye for the 1990s

"THIS WEEK"
19 August 1982

The American–European Space Telescope will observe at ultraviolet as well as optical wavelengths; but better views may come from a smaller multi-wavelength telescope to be launched a few years later.

Plans are now afoot to launch a joint American–Canadian–Australian space telescope. This 1-metre telescope, called Starlab, will "see" wavelengths from the far ultraviolet through the visible spectrum to the near infrared. Australian participation in the retrievable Starlab is a great coup for Don Mathewson, the Australian National University's Professor of Astronomy and Director of the Mount Stromlo and Siding Spring Observatory. He will have managed to project Australian astronomy into space for a modest £15 million. At the moment, NASA plans to put Starlab on the space platform the space shuttle will launch in 1989. About 10 missions, each lasting 6–12 months, are planned.

NASA has agreed to provide the platform and the first two launches at no cost to Canada and Australia. Canada will build the telescope and Australia will provide the instrument package, consisting of a camera, a spectrograph, electronic equipment for recording ultraviolet images and equipment for telemetering information to Earth. Canada and Australia will share the £30-million cost of the telescope and instrument package equally.

Mathewson's trump card has been his group's lead in the development of equipment for recording ultraviolet images – equipment which (according to Ed Weiler, chief of astronomy at NASA) puts the Australians two years ahead of their rivals. It is a photon-counting system, where the ultraviolet image is intensified with an image tube, and the optical image on the output screen is scanned and "cleaned up" by a television camera backed by a minicomputer.

16

Looking forward to X-ray satellites

KEN POUNDS

22 July 1971

The first X-ray astronomy satellite observed the sky for longer in its first day than the total time accumulated by all previous rocket flights. Its discoveries left astronomers keen for more, anticipating in particular the American High Energy Astronomy Observatory, the project which developed into the Einstein Observatory with its focusing X-ray telescopes.

The launch of Uhuru (SAS-A), the world's first satellite devoted to X-ray astronomy, took place from the Italian San Marco platform off the coast of Kenya on 12 December 1970. Placed into a 500-kilometre equatorial orbit, this small NASA satellite carries two 700 square centimetre X-ray proportional counters, designed and built by the American Science and Engineering group. The detectors are mounted on a slowly rotating platform and scan a great circle of the sky every 12 minutes; the satellite spin axis is set by ground command of an on-board magnetic torquing system. With continuous observation of a particular region of the sky for one or more days, Uhuru is now systematically surveying the whole celestial sphere with a sensitivity down to one ten thousandth of Scorpius X-1, 40 times beyond that of a typical 4-minute rocket observation. With the aid of an accurate star sensor, fixed to the side of the X-ray detectors, locations of all but the weakest sources are being obtained to a few arcminutes. Already, a significant increase in the number and variety of listed X-ray sources has been obtained, together with some quite unexpected results.

In the first category are several dozen galactic and extragalactic sources, including the bright Seyfert galaxies, NGC4151 and NGC1275, and an apparently extended source in the Coma cluster of galaxies.

The Uhuru satellite's grid-like collimator dominates this pre-launch view at the Goddard Space Flight Center, Maryland. The two lenses above are a star sensor (left) and a Sun sensor (right) which, with the collimator, helped to refine the positions of sources found by this, the first X-ray astronomy satellite

The most surprising results have been the discovery of short-period pulsations in the X-ray emission from a number of the previously known galactic sources, apparently *not* associated with radio pulsars. In Cygnus X-1 rapid pulsations occur with an amplitude up to 25 per cent of the average source intensity and a mean period of 0.29 second, while slower variability is also seen at about a 1.2-second period.

A second remarkable pulsating X-ray object seen by Uhuru is Centaurus X-3, a source observed earlier from rocket experiments by the Lawrence Radiation Laboratory and Leicester University groups. The British group found that this source varied by a factor of 10 between two Skylark flights in 1968 and 1969 and also pointed out that its peculiar X-ray spectrum is similar to that produced by a body at 23 million K. Uhuru has now confirmed both results, actually seeing a factor of 10 increase in a 1-hour period on 11 January. In addition, the extended observing period of the satellite experiment has shown that the X-ray flux is deeply modulated with a period of approximately 5 seconds. During the first few months of this year the period has also been seen to change, both increasing and decreasing by a few hundredths of a per cent in an hour or so. The rapidity of these changes is quite remarkable bearing in mind that the object involved is likely to be about as massive as the Sun. As with both the Crab Nebula X-ray pulsar and Cygnus X-1, the rapidity of the pulsations and, in this case, also the short-term variability, imply the source of radiation is very compact – probably a collapsed rotating star with a strong magnetic field.

Many other exciting and significant results can be expected from Uhuru, as successive sky scans are superimposed. The data published to date are largely derived from a single 90-minute orbit taken from one or more days of observation on a given area. It has been estimated that the total numbers of known sources may climb beyond 100 in the Milky Way and 1000 in extragalactic space. Already plans for subsequent satellite payloads are well advanced. SAS-C and UK-5, both due for launch in 1973, are small spin-stabilised satellites capable of being pointed continuously at selected sources. Plans are also well advanced for a small Japanese X-ray astronomy satellite and a joint Dutch–US payload, due for launch in 1973 and 1974 respectively.

In the American programme, a major effort in X-ray astronomy is planned for the second half of the decade. This will be based on the High Energy Astronomy Observatory, a 10-tonne Titan-launched giant now being developed by NASA. Its instruments will have a further order of magnitude in sensitivity over those of the SAS and UK-5 type and will reveal X-ray sources a few millionths of the brightness of Scorpius X-1, corresponding to the detection of such a source a thousand times farther away. This will allow the study of similar sources within other galaxies of the local group, such as the Magellanic Clouds and the Andromeda Galaxy. Still more exciting is the prospect of detecting many of the intense extragalactic sources apparently demanded by the strong X-ray background. As these sources are believed to be at the most distant reaches of space, they will provide for the first time information about the Universe in its infancy.

17

Europe's X-ray eye on space

LEN CULHANE

19 May 1983

The European X-ray observatory, Exosat, launched on 26 May, 1983, is only the second satellite to carry true X-ray telescopes. Although hampered by faults in its PSD detectors, Exosat is now returning hundreds of detailed images of cosmic X-ray sources.

X-ray astronomy began when a group at the US Naval Research Laboratory (NRL) detected X-rays from the Sun. Using a V2 rocket, launched from the White Sands Missile range in New Mexico in 1949, they discovered that an X-ray sensitive film, exposed to the Sun behind thin metal filters which did not transmit visible light, had been blackened. In the years that followed, Herbert Friedman and his colleagues at the NRL continued to observe the Sun's X-radiation with rocket-borne ionisation chambers and Geiger counters. Their work confirmed the existence of the high-temperature gas at a few million degrees that forms the solar corona, the outer region of the Sun's atmosphere beyond its visible disc.

Solar X-ray observations still continue, but astronomers now have more than the Sun to study at X-ray wavelengths. In June 1962 Riccardo Giacconi and his colleagues at American Science and Engineering Inc. achieved a major breakthrough in the course of a brief rocket flight. They discovered for the first time an X-ray source (Scorpius X-1) beyond the Solar System. Then, after almost a decade of observations with relatively simple rocket-borne instruments, the advent of X-ray astronomy satellites brought about a series of remarkable advances. NASA's Uhuru satellite, instrumented by Giacconi and his collaborators, led to the first all-sky catalogue containing dozens of bright X-ray sources.

Among the outstanding findings was the discovery that neutron stars paired in binary systems with normal stars are highly luminous

Europe's Exosat X-ray astronomy satellite undergoes tests before launch. Carried into space aboard an American Delta rocket in mid-1983, Exosat is expected to return X-ray images and spectra to Earth until early 1986

X-ray emitters. A neutron star contains matter in a very compact form, in which electrons have been "squeezed inside" protons to leave only neutrons. The matter is so dense that a mass equal to that of the Sun fits into a spherical volume of about 10 km radius. Such a star possesses an extremely large gravitational field, and attracts gaseous material to it from the ordinary companion star. As this gas is captured, or "accreted", it accelerates toward the neutron star and is compressed by the extreme gravitational field, and so is heated to a very high temperature, around a thousand million degrees (10^9 K). Most of the radiation from matter at such extreme temperatures lies in the X-ray range.

Studies of how the X-ray emission from these objects varied with time disclosed the binary nature of several of the systems and, in the cases where the X-ray source was found to pulsate, demonstrated that the neutron stars were spinning, typically once very few seconds. By combining optical and X-ray observations of these objects, it has proved possible in some cases to deduce the mass of the compact high-density X-ray source. Observations of this kind on the binary X-ray source, Cygnus X-1, indicate the presence of a compact object with a mass around 10 times that of the Sun. Such an entity could not possibly be a neutron star; instead, astronomers believe that it may be the first black hole discovered.

Another important advance followed Uhuru's discovery that large assemblies or clusters of galaxies are extended sources of X-ray emission. Later, an instrument built by the Mullard Space Science Laboratory (MSSL) of University College, London, on the British satellite Ariel V (designated UK-5 before launch), detected X-ray emission lines from iron and showed that the emission came from high-temperature gas in the space between the galaxies. Leicester University's instrument on Ariel V produced an extensive catalogue of X-ray sources which indicated that the so-called Seyfert galaxies and other active galaxies are important X-ray sources. Thus by the end of the 1970s Uhuru, Ariel V and the satellites that followed had firmly established X-ray studies at the frontiers of astronomy.

Most of the satellite instruments flown in the 1970s employed large-area X-ray detectors, with relatively crude mechanical collimation, which collected all the X-rays from a region of sky typically one degree square. Thus the angular resolution of these detectors did not match up to that of ground-based optical telescopes, which can resolve details of one arcsecond. Such simple X-ray detection systems were necessary because X-rays are extremely difficult to reflect the focus. But a foretaste of what was to come was provided by the small telescopes flown by MSSL on NASA's Copernicus satellite in 1972. Although the single reflectors of these telescopes could not form images, they provided the first detailed maps of X-ray emission from supernova remnants and of the central regions of clusters of galaxies. In addition they obtained the first evidence of time variation in the X-ray output of an extra-galactic source – the nucleus of the giant radio galaxy Centaurus A.

A great leap forward was made with the launching of NASA's Einstein Observatory in late 1978. The spacecraft contained a large imaging (doubly-reflecting) X-ray telescope provided by a consortium of institutes led by Giacconi, who had moved to the Harvard Smithsonian Center for Astrophysics. Equipped with a variety of X-ray imaging and spectroscopy systems as its focus, the Einstein Observatory revolutionised the subject during the 2½ years it was operational. For the first time, X-ray astronomers had an instrument with angular resolution comparable with that of optical telescopes. And the high angular resolution brought with it another major benefit. The telescope's detectors isolated image elements corresponding to a few arcseconds of sky and had a much smaller "background" counting rate than large-area detectors. Thus with a greatly improved sensitivity it became possible to detect objects more than a thousand times fainter than those that the crudely collimated detectors on Uhuru and Ariel V could barely register.

The Einstein Observatory has made many important discoveries. Perhaps the most far reaching of these has been the knowledge that almost all kinds of ordinary stars emit X-rays. From some points of view this was rather unexpected. Indeed early assessments of the likelihood of detecting X-rays from beyond the Solar System had been extremely pessimistic, given the knowledge of the Sun's X-ray output and the impossibility of detecting a similar source with the instrumentation available in the early 1960s, even at a distance as close as a few light years. The main achievement of the Einstein mission was to demonstrate the relevance of X-ray observations throughout the entire field of astronomy. The next step was to build on the Einstein observatory's success, and this is why European astronomers have planned the Exosat mission.

Exosat has had a chequered history. In 1970, just before NASA launched its Uhuru satellite, a group of European X-ray astronomers proposed to the European Space Agency (ESA) an imaginative technique for better establishing the positions of X-ray sources with the crudely collimated detectors that were then available. Their scheme involved the use of the Moon to obscure, or "occult", regions of the sky that contained X-ray sources. The intensity of X-rays registered by a detector crudely pointed at a source will show a dramatic decline at the instant the Moon first obscures the source. The instantaneous position of the lunar edge, or "limb", is known to better than an arcsecond, so the position of the X-ray source can be determined with comparable accuracy. A satellite – Exosat – in a highly elliptical orbit, could observe the lunar occultation of more than 100 X-ray sources in the course of a year.

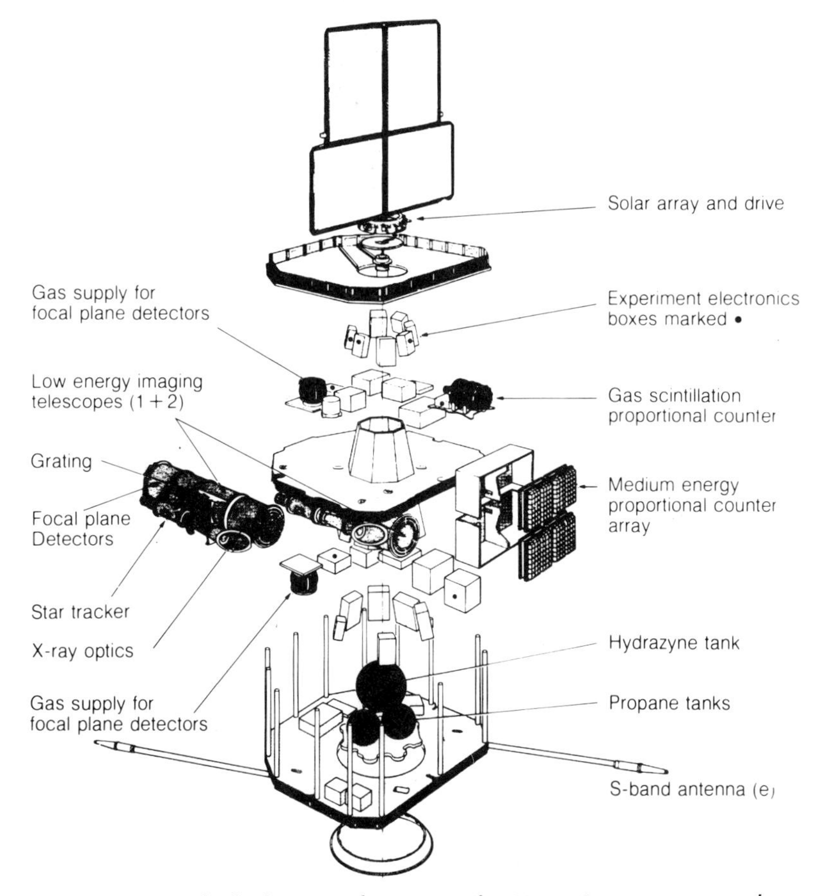

Figure 1 *Exploded view of Exosat. The X-ray instruments – the two low-energy imaging telescopes, the medium-energy proportional counter array and the gas scintillation proportional counter – all point in the same direction (to the right here)*

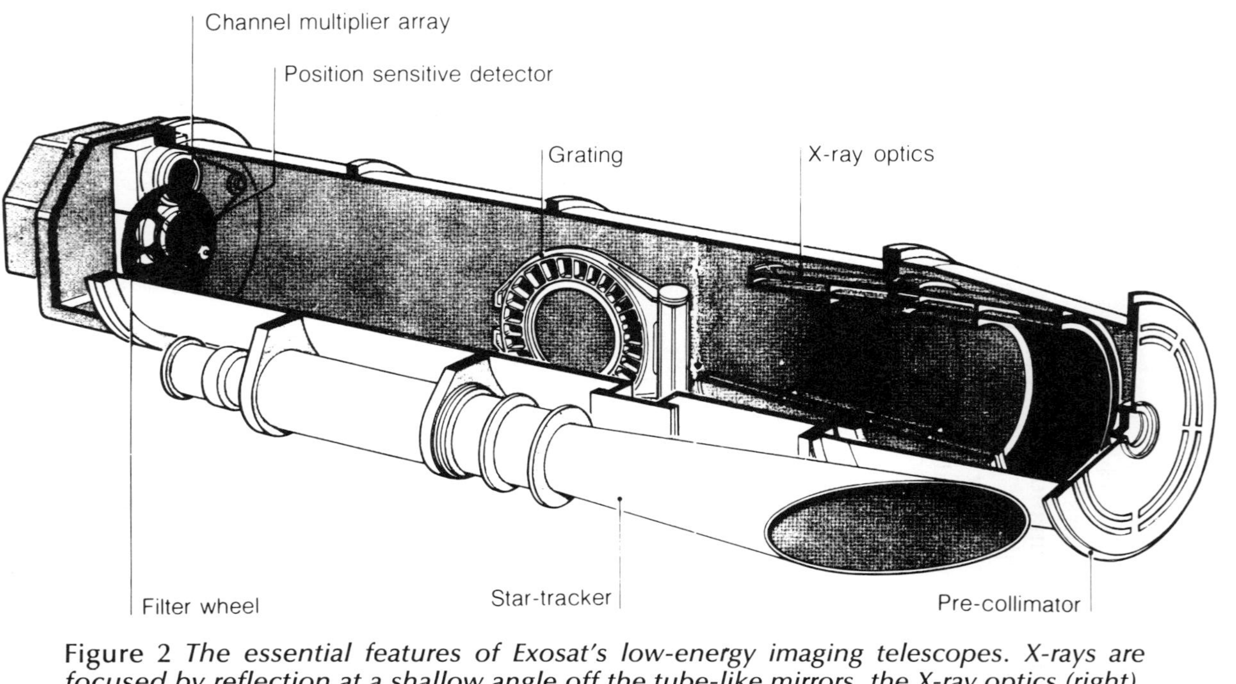

Figure 2 *The essential features of Exosat's low-energy imaging telescopes. X-rays are focused by reflection at a shallow angle off the tube-like mirrors, the X-ray optics (right), onto one of the detectors on the left. The wavelength range can be restricted by interposing filters on the filter wheel. The grating can be swung into the path of the X-rays to spread the image of a source out into a spectrum*

If it had been launched in 1977 as originally planned, such a satellite would have had a large influence on X-ray astronomy. However, for a number of reasons, but mainly because of a decision taken to launch Exosat on ESA's Ariane vehicle, the mission was seriously delayed. It soon became apparent that should the launch date slip beyond that of the Einstein Observatory, many of the proposed lunar occultation observations would be rendered irrelevant, given the ability of Einstein's telescopes to observe a whole range of sources in a short period of time. Thus the instruments on Exosat had to evolve into the satellite's present payload (Figure 1).

The original occultation payload contained a number of large-area detectors. Their purpose was to provide a sensitive monitor of the "counting rates" of X-rays from a source, so that the instant of occultation could be detected with high precision, and the position of the source determined. These detectors, which comprise the medium energy instrument, remain on the satellite. They are used for a few lunar occultation observations where it is important to search for extended halos, due for example to scattering of X-rays by interstellar dust grains around bright point sources. It is easier to detect such faint halos when the Moon blocks the flux from the central source. The medium energy instrument is also able to study the spectra and time variability of brighter X-ray sources at higher energies than the imaging telescope can respond to. The instrument was proposed by a consortium of groups from Leicester University, the Max-Planck Institute at Garching and the University of Tübingen.

There is in addition a small detector – a gas scintillation proportional counter – which operates in a similar X-ray energy range to the medium energy instrument but which has an energy resolution that is twice as good. This detector was proposed by the European Space Technology Centre's Space Science Department, MSSL and groups from the Universities of Milan and Palermo. It is used for spectral studies of brighter sources.

Two single-reflecting telescopes to provide low-energy X-ray observations of sources were part of the original payload. These have been replaced by a nested pair of doubly reflecting imaging telescopes of the Wolter I design (Figure 2). The telescopes are capable of an angular resolution in the region of 10 arcseconds. In each telescope, radiation reflected from the second (hyperboloidal) surface is brought to a focus on the face of one of two interchangeable detectors: a position-sensitive proportional detector (PSD) and a 24-channel multiplier array (CMA). Both of these systems can register the position of an arriving "packet" or photon of X-ray radiation. The spatial resolution of the PSD is poorer than that of the telescopes; with this detector at the focus the overall angular resolution is about an arcminute. However, the signal from the detector is proportional to the energy of the incoming X-ray photon, so it provides information about the X-ray spectrum of the source.

With the CMA at the focal plane the telescope's angular resolution of 10 arcseconds can be realised, but the signal from the detector contains no information about X-ray energy. Instead, a variety of filters with differing transmission bands may be inserted in the beam in turn, thus providing some crude spectral information. For higher spectral resolution, transmission X-ray diffraction gratings can be placed in the beam behind the telescopes. Gratings of 500 and 1000 lines/mm are available. When used with the CMA detectors they provide higher spectral resolution than can be achieved with the PSDs. The two imaging telescopes were proposed by a consortium that included the MSSL and two groups from Leiden University and the Utrecht Space Research Institute. Taken together, the telescopes have a collecting area half that of the Einstein Observatory.

Control of the satellite is carried out from ESA's operations control centre at Darmstadt in West Germany. An antenna located at Villafranca near Madrid and land-linked with Darmstadt receives data from the satellite and transmits commands to it. These commands not only contain instructions on where the telescopes are to point, but also on which detectors, filters and so on are to be positioned at the focal plane.

ESA's Council decided on 22 February 1983 that Exosat would be launched not by an Ariane rocket but on a NASA McDonnell Douglas Delta 3914 vehicle from Vandenberg Air Force Base in California. The change was a consequence of the rescheduling of the Ariane launching programme after the fifth Ariane rocket failed, in September 1982. A series of communications satellites would instead be launched on Ariane vehicles during the remainder of 1983. Launch on a Delta vehicle allows Exosat to be flown before the closure of the summer launch window, keeping down the total cost and ensuring a launch before problems about the storage lifetime of the detectors would have to be faced. The first six weeks following the launch are devoted to calibrating and commissioning the satellite. During this time calibrations of the instruments made before the launch can be verified by observations of well-known

X-ray sources, such as the Crab Nebula whose intensity is well established.

ESA received more than 500 replies to its call for proposals for observations to be undertaken in the first nine months of Exosat's life. They contain ideas relevant to the entire range of modern astrophysics. This response represented more than 10 times the amount of work the satellite can do – a testimony to both the importance of X-ray astronomy and to the lack of any other X-ray instruments (apart from two small Japanese satellites) in orbit at the present time. With a planned minimum lifetime of two years and a good probability of almost doubling this figure in practice, it is clear that given a successful launch and operation in orbit, the Exosat mission will establish Europe's X-ray astronomers in a position of world leadership for the rest of the decade.

18

X-ray astronomy in British, German and Japanese hands

"THIS WEEK"

18 March and 21 October 1982 and 13 January 1983

America's hopes in X-ray astronomy are pinned on the huge and expensive Advanced X-ray Astrophysics Facility, which has yet to be funded. For the next few years the major incentives will come from West Germany, Britain and Japan, with the USA collaborating by providing some detectors and – in the case of Rosat – a free launch on the space shuttle in return for half the satellite's observing time.

Exosat will be followed by the first true successor to the Einstein Observatory. This satellite will be neither American nor European – but a West German national project. This X-ray observatory is Rosat, an abbreviation of *Röntgen-Satellit* – the Germans refer to X-rays as Röntgen rays, in honour of the German physicist Wilhelm Röntgen who discovered this type of radiation in 1895. Due for launch in early 1987, the 2-tonne Rosat will carry a larger version of the Wolter I telescope used on Einstein. With an outer diameter of 80 centimetres, and more modern detectors, Rosat will be three times more sensitive than Einstein, and will have roughly the same ability to see detail.

For its first six to nine months, Rosat will not investigate individual objects; instead, it will scan the entire sky to compile a complete catalogue of X-ray sources. This should reach down to sources a hundred times fainter than the previous most sensitive survey, from HEAO-1, and should net a total of some 100 000 X-ray sources in all. For the next 18 months, to the end of its design life, Rosat will follow up its new discoveries by looking at some 10 000 individual X-ray sources in detail.

Britain will be a minor partner in Rosat. The Germans invited other members of the European Space Agency to participate in Rosat, and here Britain's long experience in X-ray astronomy paid

The X-ray imaging mirrors aboard the German Rosat – due for launch by space shuttle in 1987 – will be slightly larger than this mirror being prepared for the Einstein Observatory in 1977. X-rays are reflected off the inside of the cylindrical surface at a very shallow angle

off. The Germans decided that the proposal "most complementary to the main mission" would be a British "wide-field camera". This will be second X-ray telescope, designed to observe longer wavelengths: the main telescope on Rosat will detect X-rays between 0.6 and 5 nanometres, while the British "camera" is designed for the range 5–30 nanometres.

The British contribution to the mission will cost £8.75 million – partly for the camera itself, and partly as payment to the Germans for providing extra electric power and to compensate for the increased cost of launching a slightly larger satellite on the space shuttle. The construction of the camera will involve all Britain's X-ray astronomers. The major groups at Leicester University and the Mullard Space Science Laboratory (of University College,

London) will use their experience in building the detectors, electronics and mirror assembly. The nested mirrors themselves will be made on a new turning machine now undergoing acceptance trials at the Cranfield Unit for Precision Engineering. Birmingham University will build the support structures, and Imperial College the charged particle "background" monitors. The Rutherford Appleton Laboratory will handle a number of tasks, including the construction of star sensors to check the camera's pointing position; it will also take on the sensitive diplomacies of international collaboration.

The British camera on Rosat will be firmly fixed to the German telescope, so both will always point in the same direction, although the British wide-field camera will "see" an area about 10 times that of the German telescope's field of view. During the initial part of the Rosat mission, the British camera will scan the sky to produce a survey at longer wavelengths than the German instrument. According to Ken Pounds, principal investigator for the British camera project, the survey will be the first ever at these wavelengths, which span the borders of the X-ray and the ultraviolet. Astronomers commonly refer to such wavelengths as XUV (X-ray-ultraviolet) or EUV (extreme ultraviolet). The survey is thus bound to come up with some surprises. In the second, "pointed" phase, British astronomers will be able to choose which target to observe for 15 per cent of the observing time – in line with Britain's financial contribution to Rosat.

British astronomers are also to provide the main X-ray detector for a new Japanese astronomy satellite, ASTRO-C, which should be launched on a Japanese rocket in 1987. The collaboration means that Britain will be the only country with experiments on board the three major X-ray satellites now scheduled for this decade – Exosat, Rosat and ASTRO-C.

The Japanese have picked the British detector because of the success of earlier British X-ray satellites such as Copernicus and Ariel V. Japan intends to be the West's leading nation in the scientific use of space by the late 1990s. The British, meanwhile, see this latest Anglo-Japanese collaboration as further evidence of a growing interchange of scientific research on a wide front.

The new X-ray detector, one of the largest ever constructed, will be built jointly by astronomers at the University of Leicester and the Mullard Space Science Laboratory of University College, London. It will not be designed to produce focused pictures. Instead it will be very sensitive to changes in the X-ray brightness of sources. It will also measure spectra in the wavelength range from 0.02 to 1 nanometres.

There may also be a British-led X-ray satellite. This £50 million satellite could be launched by the end of the decade as part of the current reincarnation of the British space programme.

The satellite would not only "see" thousands of faint X-ray sources but would also be sensitive enough to examine their spectra. This would, in turn allow scientists to analyse the temperature and composition of the gases in deep space that produce the X-rays.

The principal instrument of the X-ray satellite is to be the High-Throughput Spectrometer. This consists of a traditional X-ray telescope, comprising a nest of seven cylindrical mirrors, one inside the other. The largest, outer cylinder will measure 1 metre across. It is considerably larger than the largest X-ray telescope so far launched. The Einstein Observatory, which took off in 1978, had only four nested mirrors, the largest being 0.6 metre in diameter.

The High-Throughput Spectrometer will record the spectra of X-rays by splitting up the rays into bands of differing wavelengths using a "venetian blind" of gratings in front of the mirror assembly.

The second telescope on the satellite will obtain the first images and spectra at very short X-ray wavelengths. This high-energy telescope is a "tinfoil telescope", using curved sheets of aluminium foil and was pioneered at America's Goddard Space Flight Center.

Other countries may provide some of the instruments in return for booking observation time. The Americans could provide a free launch aboard the shuttle. This alone would save £15 million.

19

The gamma ray sky

MICHAEL HILLAS

28 April 1977

One of the earliest European astronomy satellites, COS-B, revealed the appearance of the sky when "seen" at the shortest wavelengths of all. Its "telescope" was more reminiscent of a physicist's particle detector than anything seen in a conventional observatory, and even the preliminary results provoked controversy among astrophysicists.

Radio astronomers have found that at the very longest wavelengths the Sun no longer dominates the sky to the extent it does optically, and a large amount of radiation comes from the regions between the stars in our Galaxy. The stars themselves are not seen, and pulsars and radio galaxies make their appearance. In the last few years, the first maps of the sky have been drawn at the other extreme of the electromagnetic spectrum, using very high-energy gamma rays. The Sun and stars do not show up, and, perhaps rather surprisingly, the pictures at the two extreme ends of the spectrum look rather similar – at least within our own Galaxy. It seems that in both cases we are studying the non-thermal radiation of the Galaxy; in between, at visible wavelengths and in X-rays, we generally see thermal radiation, due to stars and very hot gases.

Attention was directed to what one might see in other parts of the spectrum as soon as the potential of radio astronomy was recognised, but it has been much more difficult to pursue such observations; for whereas the radio waves reach us through an atmospheric "spectral window", a region of wavelengths to which the air is transparent, gamma rays are absorbed in the upper atmosphere. Not only that, but far more gamma rays are generated in the atmosphere under the bombardment of cosmic-ray particles. So the gamma ray universe is indecipherable until one gets almost outside the atmosphere by means of high-altitude balloons or satellites.

The gamma ray satellite COS-B just prior to launch in 1975. The 1.4-metre diameter cylinder contained just one experiment, designed to detect and pinpoint the locations and energies of gamma ray sources within our Galaxy. In a 6½-year lifetime it detected over two dozen sources, most of which are still unidentified

Although the first positive detection of gamma rays from the Galaxy was made by the OSO-3 satellite in 1968, very variable results emerged from the early balloon experiments. Only recently has the position become clear as the result of observations made from two satellites devoted entirely to gamma ray astronomy, which have given an excellent map of much of the sky seen in the light of 100-MeV gamma rays. The first was the Small Astronomy Satellite 2 (SAS-2), developed by NASA's Goddard Space Flight Center, and second the European COS-B satellite. It now seems that gamma ray astronomy will contribute substantially to answering questions on the origin of cosmic rays; on the large-scale structure of our Galaxy (much of which is hidden from optical view by dust clouds); on the existence of a very extensive "halo" around our Galaxy; on the mechanism by which pulsars pulse; and perhaps may give another view of the early evolution of the Universe.

The link between gamma ray astronomy and radio astronomy is through cosmic rays. The term "cosmic rays" is normally used to refer to charged particles moving at relativistic velocities – mostly protons, with other atomic nuclei and a few electrons – which reach us from all directions in space. The origin of their very high individual energies, and the large energy flux carried altogether by these particles, presents one of the long-standing problems of astrophysics.

A principal obstacle to tracing the origin of cosmic rays is the erratic deflection the particles suffer in traversing magnetic fields in space. When the relativistic electrons are deflected in magnetic fields, though, they emit synchrotron radiation, normally at radio wavelengths, which has allowed radio astronomers to map out the presence of cosmic ray electrons in distant regions. This technique has shown that cosmic ray electrons are spread throughout the Galaxy at least, have local concentrations in special places such as the Crab Nebula and other supernova remnants, and appear in some very intense extragalactic sources. However, there are two reasons why we may be misled by the radio observations: radio emission depends on magnetic field strength as well as electron concentration, and the field strengths have been deduced only indirectly; and it is only the electrons, which form a very minor and possibly unrepresentative proportion of the cosmic rays, that are accelerated sufficiently to radiate in this way.

G. W. Hutchinson of Southampton University, Philip Morrison of MIT, V. L. Ginzburg of the P. N. Lebedev Physical Institute, Moscow, and others had pointed out that cosmic ray particles should generate gamma rays as they passed through the gas normally present in interstellar space. Protons, in particular, should occasionally collide with atomic nuclei to produce pi-mesons, some of which decay into gamma rays of typical energy around 100 MeV and upwards. Hence, as we have a good idea of the density of atomic hydrogen from 21-centimetre radio observations, we should be able to trace the passage of relativistic particles. Relativistic electrons, too, can emit gamma rays (by the process of braking radiation or *bremsstrahlung*) in the gas, and may in some places emit gamma rays significantly by colliding with low-energy photons from starlight or other sources. It seems that for gamma rays below 50 MeV, electron-generated gamma rays should predominate, and above 100 MeV protons should be the main source. Measurement of the gamma ray energy spectrum should confirm the source mechanism.

The pioneering observations were made with a simple shielded scintillator detector having a low angular resolution. But the greatly developed experiments in the SAS-2 and COS-B satellites both used spark chambers, capable of showing the direction of arrival of each gamma ray with some accuracy. The apparatus illustrated in Figure 1 is the system flown in COS-B, launched from NASA's Western Test Range in August 1975. It was developed by the European Space Agency, based at Noordwijk, in collaboration with groups at Munich, Saclay, Leiden, Milan, Palermo and, in the early stages, Southampton. The American SAS-2 satellite, which operated from November 1972 until a failure in June 1973, had a similar detector.

An energetic gamma ray (of energy tens of MeV or more) would usually penetrate the outer domed scintillation counter A without producing any signal, and would normally create a positron and electron in one of the thin tungsten plates of the stack of spark chambers S, the particles travelling forward close to the direction of the incoming photon, but then gradually moving apart. The Cerenkov counter C and scintillator B detect the pair of particles emerging and cause a high-voltage pulse to be applied to alternate plates of the stack, so that lines of sparks mark out the trajectories of the electron and positron. (In order to pick out the few gamma rays

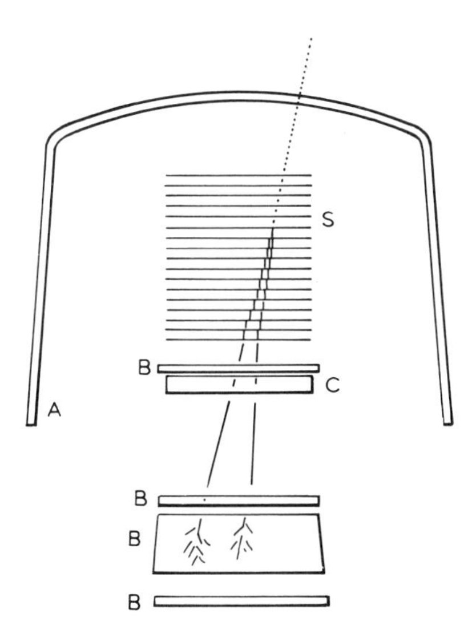

Figure 1 *The detector aboard the COS-B satellite. An incoming gamma ray photon (dotted line) converts to a positron–electron pair in one of the thin tungsten plates in the stack S, and lines of sparks mark their trajectories. A and B are scintillators, and C is a Cerenkov counter sensitive only to "downward"-going particles*

from the overwhelming flux of charged particles traversing the detector, the chamber is not pulsed if scintillator A detects a charged particle entering, and C is blind to "upward"-moving particles.)

The sparks are picked up on wire grids, and their coordinates are transmitted back to ground. The direction of approach of the incoming photon is deduced as being intermediate between the tracks of the two charged particles; above 100 MeV, the accuracy is typically around 2–4 degrees, but is not so good at lower energies. The total energy deposited in a thick scintillator, at the bottom of the stack, gives the energy of the photon. Then, from an examination of the pictures, those records which do not appear to involve pair production can be eliminated, a rough energy spectrum of the rest can be obtained, and a map of their arrival directions plotted. Before the apparatus was flown, its performance was checked with gamma rays of known energy and direction at the DESY accelerator at Hamburg.

At energies above 100 MeV the Sun is a negligible source. The strongest emission comes from a narrow belt stretching about 40 degrees either side of the centre of the Galaxy along the Milky Way, and the narrowness of the belt suggests that the radiation comes from interactions in the thin sheet of interstellar gas. The pattern is that which would be observed if the emission were peaked in an annulus of about 15 000–20 000 light years radius around the centre (we are situated about 30 000 light years from the galactic centre), together with a peak at the centre of the Galaxy. But the variation in density of atomic hydrogen, as deduced from 21-centimetre radio emission, does not agree with the derived variation of gamma ray emission with distance from the galactic centre. It appears at first as though the flux of cosmic ray protons must be very much higher in the region of this peak than near the Earth, if we are indeed to ascribe these gamma rays to nuclear collisions in the gas (and their energy spectrum supports this interpretation).

The situation has turned out to be less clear-cut. The sudden fall-off in emission near 20 000 light years suggested that there are previously unrecognised belts of gas clouds in this region of the Galaxy. Around the time that this result emerged, the radio astronomers P. M. Solomon and N. Z. Scoville interpreted radio observations of the 2.6-millimetre carbon monoxide line as indirect evidence for the existence of a belt of patchy clouds of molecular hydrogen in this zone, the hydrogen molecules being inferred in order to excite the carbon monoxide molecules by collision. The gamma ray evidence was welcomed as supporting data.

Until maps of much higher angular resolution are available, we shall not be certain how large a part these dense gas clouds play in the production of gamma rays, but the best estimate at present leads Floyd Stecker of the Goddard Center to the conclusion that the density of cosmic ray protons does indeed rise as one goes towards the central region of the Galaxy in the same way as the frequency of supernova explosions, lending support to the view that cosmic rays originate from supernova outbursts.

The position is much clearer if one looks away from the galactic centre. Near the anticentre direction the intensity is much lower. Using the 21-centimetre radio observations again to deduce the amount of atomic hydrogen we look through, Arnold Wolfendale and his collaborators at Durham University have pointed out that, even in the absence of any other gas (for example, hydrogen molecules), the observed intensity of gamma rays is much lower than we would expect if the cosmic ray flux through this gas were as large as that found near the Earth. So cosmic rays are less numerous in the outer fringes of the Galaxy, supporting the idea that most of the cosmic rays are indeed produced in our own Galaxy, rather than outside.

A few localised sources stand out from this general pattern. The two strongest have been identified with two of the youngest pulsars – in the Crab Nebula, and in the Vela supernova remnant – and the gamma rays have been found to emerge in sharp pulses with the correct period. In the Crab this is not surprising, as the pulses had been traced at radio, optical and X-ray wavelengths, but the Vela pulsar had not then been detected optically or in X-rays; and the gamma ray pulses are out of phase with the radio pulse, but yet are very sharp. It looks as though the emissions at different wavelengths originate at different points along the magnetic field lines attached to the rotating pulsars, but more cases will be needed to establish any pulsar model. The Goddard group believe they have detected two other pulsars which radiate much of their energy in the gamma ray region, and believe also that they may have periodic emission from the X-ray binary Cygnus X-3, which up to now had been considered to be a thermal radiation source. Evidently we are due for some surprises.

Not all cosmic gamma rays come from the Galaxy. On the Apollo 15 and 16 flights to the Moon, a scintillation counter was deployed outside the spacecraft to measure the spectrum of lower energy gamma rays (below 30 MeV) in space, and the result proved very perplexing. Superimposed on a spectrum which falls smoothly with increasing energy there appeared a considerable excess of photons in the 1–30-MeV range, a result reported also by balloon experiments. In this energy range the flux seemed more or less the same in every direction, and hence seemed to originate well beyond our Galaxy. There has been much argument about these results because of disturbing effects of nuclear reactions (in the atmosphere for the balloon experiments, and in the detector for the Apollo experiment), so judgement on the theoretical significance of these spectra has been suspended; but not before some exciting possibilities of gamma ray astronomy had emerged.

Stecker and collaborators examined possible production mechanisms for gamma rays in this range, and concluded that if the gamma rays came from decay of pi-mesons they must be enormously red shifted to bring the main emission to the 1-MeV region (rather than 70–100 MeV), and hence originated during the very early history of the Universe. One of the most promising explanations favoured a cosmological model in which the Universe began with equal amounts of matter and antimatter mixed together; this generated gamma rays by mutual annihilation until only large islands of matter and antimatter were left. However, we are far from being able to draw such conclusions yet. Quite apart from the dubious accuracy of calculations of the detailed history of such a complicated Universe, much of the radiation may turn out to originate in numerous distant Seyfert galaxies, for instance, or even to a large extent in a large halo which cosmic ray physicists believe to surround our Galaxy.

20

Masks sharpen the gamma ray view

MALCOLM COE

13 October 1983

Future gamma ray telescopes will provide detailed views of the still-mysterious gamma ray sources.

If we look critically at the pioneering results from the first generation of gamma ray telescopes, we can identify two areas of concern. Why have these telescopes left unanswered many questions about the nature of the gamma ray background? And in what direction should the design of the next generation of telescopes move in order to answer these questions?

The main problem to be tackled is the nature of the unidentified point sources. The COS-B satellite, which discovered most of them, was unable to say exactly where these sources were located. The uncertainty is one to two degrees, and within such an area of sky one may expect to find dozens of optical objects and many radio and X-ray sources. COS-B could not even say whether the sources were truly point-like or extended objects up to a degree in size. Clearly, if we are to make progress in identifying these gamma ray sources and understanding their true nature we must first find out exactly where they are, and how big they are. Thus, the first and main objective of any new telescope has to be a much-improved angular resolution.

Extending the range of photon energies over which we observe them will also make the identification of gamma ray sources easier. COS-B operated over the photon energy range 70 to 5000 MeV, and most existing X-ray measurements stop at energies below 0.1 MeV. If we could bridge this spectral gap we might well be able to match the spectra of these gamma ray sources with the established spectra of X-ray sources. As a result we will learn a great deal more about the energy output of these sources, and we will also have a much better chance of identifying them with known X-ray sources. Thus, any new detector should also extend the spectral range over which these objects are observed.

Finally, if we simply make our telescope larger (a perennial cry from astronomers at any wavelength!) we will make it more sensitive, with two consequent advantages. We will increase the time-resolution with which we can study these objects – the ability to study rapid changes in them – and this will allow us to look for periodicities due to a neutron star (pulsar) rotating or moving around a companion star. In addition, we would be able to see many more sources, and could thus begin to answer the question of whether or not the diffuse emission of gamma rays from our Galaxy is simply the combined result of many faint sources, so far unresolved.

Having thus established our guidelines for future gamma ray telescopes, let us look at proposed experiments for the next decade. There is a good chance that the US will launch its American Gamma Ray Observatory in the late 1980s. This satellite consists of a mixture of traditional gamma ray instruments, as flown on previous satellites and on balloons. It certainly satisfies our second and third requirements for a "next generation telescope" in that it both greatly extends the spectral range of the observations and is larger than previous instruments. However, its ability to locate the positions of sources is only marginally better than that of COS-B. Thus, we are left with the possibility that the observatory may discover many more sources and measure their spectra but the crucial lack of an accurate position in the sky may prevent astronomers identifying them. On the other hand, the much more detailed information that will be available on the sources' spectral and temporal variability should greatly enhance our understanding of them.

In many ways the Gamma Ray Observatory will represent the best one can do with traditional instrumentation, so several European institutes are working on a totally new approach to gamma ray astronomy. What is required is a telescope that will actually produce gamma ray images of the sky, with a resolution as good as an arcminute. Unfortunately, no device can image gamma rays in the normal, optical, sense. They cannot be focused with lenses or reflected with mirrors, since they pass straight through such objects. But it is possible to create a sky image using the technique of coded masks (Figure 1).

A coded-mask telescope consists essentially of two elements: a binary mask made of lead or tungsten squares arranged rather like a cross-word puzzle; and a position-sensitive detector plane, a two

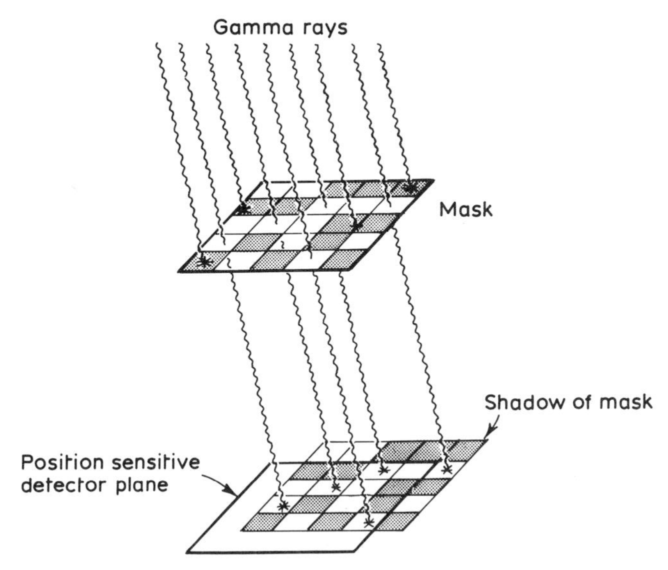

Figure 1 *In a coded-mask telescope, the position-sensitive detector plane is used to measure the position of the mask's shadow, and hence ascertain the direction from which the gamma ray photons are coming*

dimensional detector, which can identify the point at which the gamma ray photon makes contact. For low-energy gamma rays the detector consists of caesium iodide or sodium iodide crystals viewed by many photomultiplier tubes to detect the flash of light produced by a gamma ray photon. If we set up this instrument with a suitable gap between the mask and the detector plane, and point it at the sky, then any source of gamma rays will cast a shadow of the mask onto the detector plane. The detector can locate the position of the patterned shadow and, knowing the distance to the mask, we can determine the position of the source in the sky. The mask pattern is chosen in such a way that the shadow of any part of it cannot be confused with that from any other part of the mask. In fact, this technique is so powerful that we can use it to study simultaneously all sources within the field of view of the instrument as well as the regions of the sky between the sources where we detect the diffuse cosmic emission. A typical field of view for such a telescope could be 10 degrees square, and within that region of the sky any source detected could be located to an accuracy of about 5 arcminutes. In addition, the device simultaneously provides data on the gamma ray spectrum of every point in its field of view, for both sources and background.

For the lower-energy end of the gamma ray spectrum, the British and Italian research councils have funded a balloon-borne coded-mask telescope which will, it is hoped, operate from 1985 onwards. This device (known as the Zebra telescope, from the striped nature of its crystal detector plane) will cover the important gap in the gamma ray spectrum from 0.1 MeV to 5 MeV. Southampton University in conjunction with research laboratories in Milan, Bologna and Frascati are building the telescope which should produce images with a resolution of about 10 arcminutes. The European Space Agency is considering for a future satellite mission a similar device, but with even greater imaging capabilities.

For gamma rays with higher energies (above 20 MeV) the crystal detector plane could be replaced by a spark chamber (the kind of detector used, without a mask, on COS-B), or with a drift chamber. The latter has been gradually replacing the spark chamber in high-energy physics experiments in recent years, because it is capable of much greater accuracy in locating the paths of the electron and positron, and hence in revealing the point at which the incoming photon first hit the detector plane. Apart from the detector, such high-energy telescopes are similar to the low-energy version discussed above. A Soviet-French satellite, Gamma-1, will carry a spark chamber version into orbit in a few years' time, while a drift chamber telescope is being developed in Britain, in conjunction with French and American colleagues.

Many gamma ray astronomers now believe the next significant step forward in this young and exciting branch of astronomy lies with the new coded-mask system. It meets, and surpasses, the three crucial guidelines for a "next generation" gamma ray telescope, and represents an opportunity for European astronomers to capitalise on the tremendous pioneering work of COS-B. With such a device Europe could grasp the lead in gamma ray astronomy and fling open our final window on the Universe.

PART FIVE

New Frontiers in the Infrared

The wavelengths between visible light and radio waves both constitute the oldest branch of non-optical astronomy, and are also one of the main frontiers of modern research. ''Infrared'' is a general term for this very broad wavelength band, which ranges from the red end of the optical spectrum at 0.7 micrometres (700 nanometres) to the arbitrary boundary with radio waves at 1000 micrometres (1 millimetre). The waveband is sometimes broken down into the near infrared (0.7–4 micrometres), middle infrared (4–40 micrometres) and far infrared (40–300 micrometres). The wavelengths 300–1000 micrometres (0.3–1 millimetre) are now called *submillimetre waves*, because they are studied by techniques similar to those that the radio astronomers use for wavelengths of 1–10 millimetres. (For convenience, the latter are also included here.)

Infrared was the first ''invisible radiation'' to be discovered, when the astronomer William Herschel found in 1800 that a thermometer would register heat when placed beyond the red end of the Sun's visible spectrum. This detection, the first in infrared astronomy, was a year before Johann Ritter discovered ultraviolet radiation – again from the Sun – and almost a century before radio waves (1888), X-rays (1895) and gamma rays (1900) were produced and identified in the laboratory. The astronomies of the latter four wavelength regions came much later. By the late 1970s, however, radio, X-ray and ultraviolet astronomy were far more advanced than infrared astronomy, with instruments such as the Cambridge Five-Kilometre and the American Very Large Array radio telescopes, and the International Ultraviolet Explorer and the X-ray Einstein Observatory satellites.

Infrared astronomy has been restrained by three main problems. The first is simply that sensitive detectors were not available until the development of semiconductor materials in the 1960s and 1970s. These ''half-metals'' include silicon (used in ordinary semiconductor

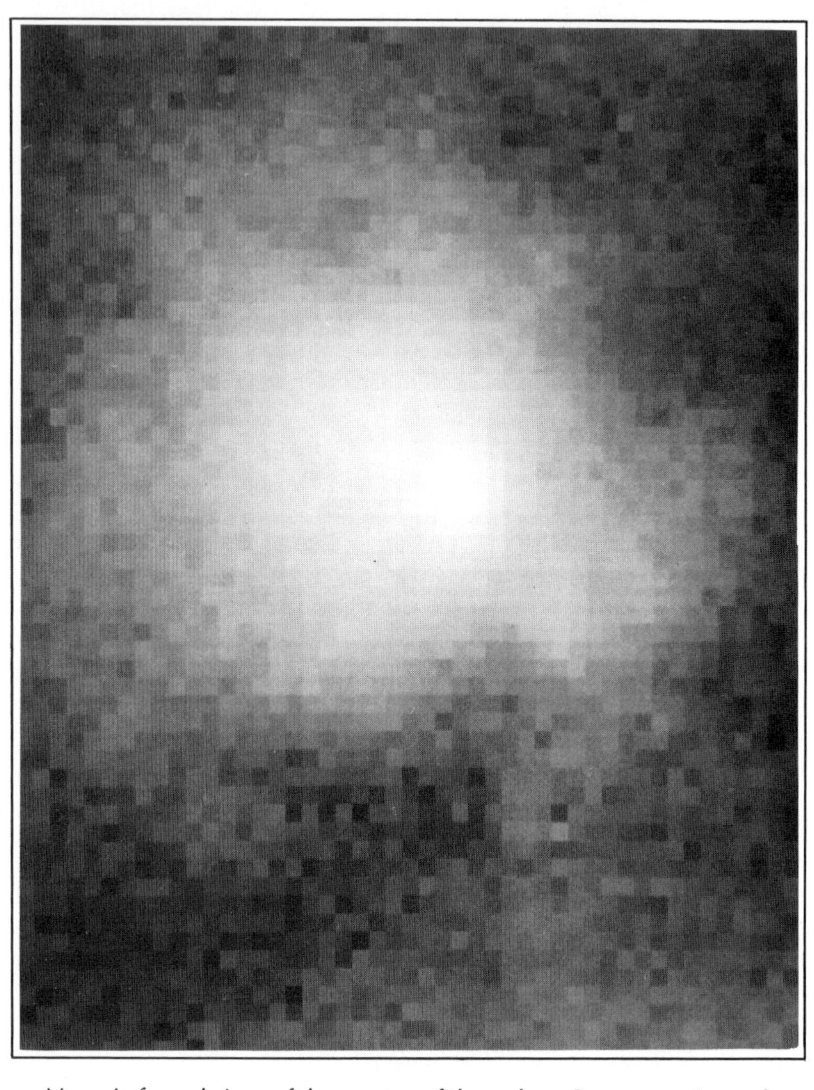

Near-infrared view of the centre of the galaxy Centaurus A, made with the Anglo-Australian Telescope, shows the high concentration of stars there. This dense core is completely hidden from optical astronomers by dust clouds, which absorb light – but not infrared

chips), indium and germanium, usually "doped" with small amounts of gallium, bismuth, beryllium or antimony. For the infrared wavelengths the detectors respond to the intensity of the radiation in very much the same way as a photographic exposure meter. In the submillimetre range the electromagnetic waves are made to produce a very high frequency alternating electric current, in a similar manner to the way a radio telescope operates. To add to the complexity, the semiconductor detectors must be cooled. For some semiconductors, 80 K will suffice, but others must be chilled right down to 3 K with superfluid liquid helium.

A second problem afflicts the middle infrared wavelengths. Objects at room temperature (300 K) emit their natural black body radiation at wavelengths of 5–50 micrometres, with a maximum at 10 micrometres. The atmosphere is at this temperature, and thus emits a permanent middle infrared glow; and so does the telescope. The problem is similar to that facing an optical astronomer who looks at stars in the daytime, through a luminous telescope! Careful design helps to prevent stray radiation from the telescope reaching the detector. The easiest way to circumvent the radiation from the sky is to observe alternately the source (including radiation from the surrounding sky) and an equal area of sky alone: the difference between the two signals is then the radiation from the source.

The third enemy of the infrared astronomer is again the atmosphere, but in a different role. Carbon dioxide and water vapour in the air absorb infrared radiation. Because these gases occur primarily in the lower atmosphere, infrared and submillimetre wave astronomers have sited their ground-based telescopes on mountain peaks – the summit of Hawaii, at an altitude of 4200 metres, being particularly suitable. From sites like this, astronomers have several "windows" (narrow wavelength bands where the air is relatively transparent) at near and middle infrared wavelengths; and, when the air is dry, at submillimetre wavelengths too. But the far infrared, 40–300 micrometres, cannot be observed at all from the ground.

Observations have been made from aircraft, balloons and rocket flights, but the obvious (and expensive) answer is a satellite. Out in orbit about the Earth, a satellite escapes all the problems associated with the atmosphere, and if the telescope is cooled down from room temperature to a few degrees K, astronomers can escape the problem of looking through a "luminous telescope". The technical difficulties are formidable, and the first infrared satellite did not fly in until 1983. This highly successful survey instrument, IRAS, will be followed by cooled infrared telescopes carried on the space shuttle's manned

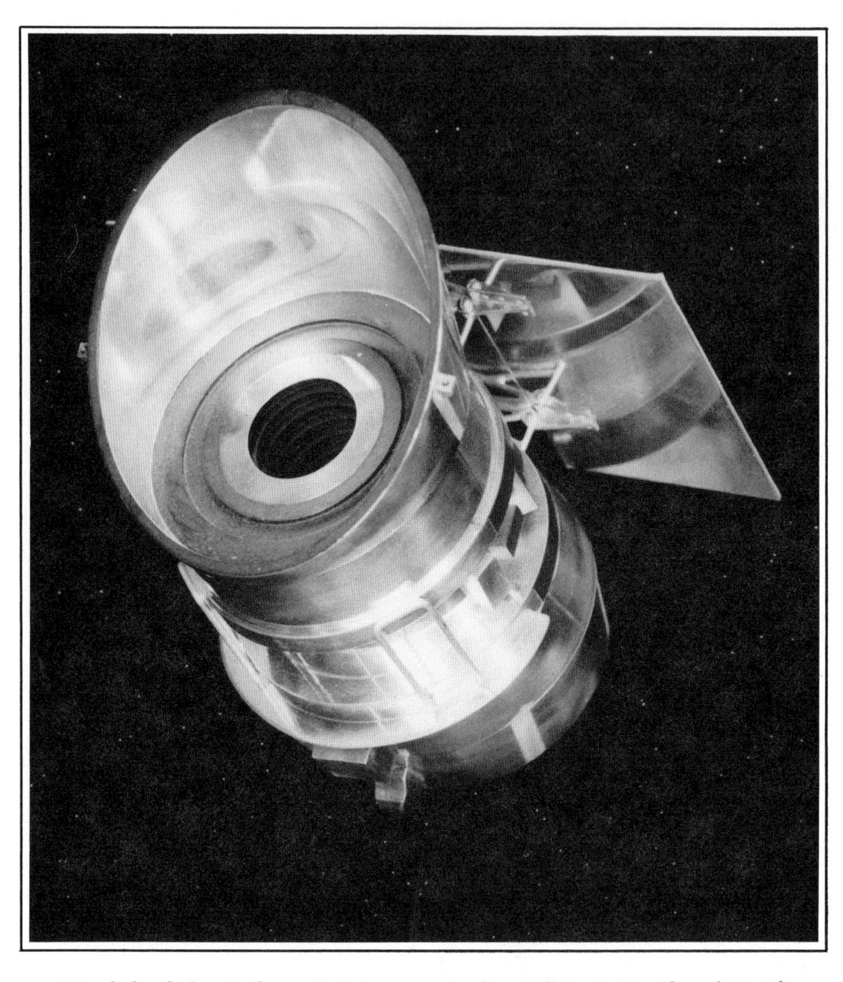

Model of the Infrared Astronomical satellite in Earth orbit. The aperture through which radiation enters is visible at the top. It is surrounded by a thick cooling jacket filled with liquid helium, which prevents the satellite from emitting its own heat radiation

laboratory, Spacelab, and by a European – and possibly an American – infrared observatory later in the 1980s. They will investigate in detail the structure, spectra and polarisation of sources found by IRAS.

These infrared sources are mainly regions of dust at temperatures of 30–300 K. In our Galaxy, such dust is either found in dense clouds in regions of star formation, or is ejected from old stars. Distant galaxies and quasars can radiate a hundred times as much power in the infrared as they do at optical wavelengths, with the radiation coming either from extensive dust clouds or from an active centre, probably gas streaming into a massive black hole.

The excitement of submillimetre wave astronomy lies in investigating the spectral lines that are found in multitudes at these wavelengths. The radiation in these lines comes from molecules in space, usually located inside the dense dust clouds "seen" by infrared telescopes. The molecular lines can, however, tell us of regions cooler than infrared astronomers can detect, and the doppler shifts in their wavelengths can reveal the motions of this gas.

Millimetre and submillimetre wave astronomy have also opened the entirely new field of *astrochemistry*. Information on the various types of molecule found in space indicates the course of reactions that occur at such low temperatures, densities and pressures that they take thousands or millions of years to come to equilibrium, and so cannot be investigated in the laboratory. Telescopes for these wavelengths are now being built on several mountain tops around the world, and in a decade's time we can expect to see submillimetre wave telescopes flying on satellites too – thus plugging the long-standing gap in our knowledge of the sky between the atmosphere's optical and radio windows.

21

The largest infrared telescope on Earth

MICHAEL MAUNDER

6 April 1978

The UK Infrared Telescope (UKIRT) is the largest of its kind in the world, and pioneered the use of very thin – and hence lightweight – mirrors. Astronomers have unique problems in using a telescope at the altitude of 4200 metres, but UKIRT has fully lived up to the expectations voiced in its early days.

A "trend setter" is probably the correct way to describe the new 3.8-metre British infrared telescope on Mauna Kea, Hawaii. The Hawaiian site is accepted as being the best in the world, with excellent seeing for optical astronomy, and has the added attraction for infrared observation of being well above the main weather layers and the bulk of the atmospheric water vapour which is largely opaque to infrared waves.

Infrared observers can operate in daylight, and on about 70 per cent of the nights; they can also detect radiation through dust. A 1.5-metre prototype instrument on Tenerife readily penetrated dust in the upper atmosphere from the Sahara, giving a rather mundane advantage over optical telescopes at the same site. The ability to observe through cosmic or interstellar dust, however, opens up new avenues of research. Very dense interstellar dust clouds seem to be a major factor in the production of infrared stars – those which are bright in the infrared but very faint in visible light. The dust obscures light from the region behind it, but allows infrared radiation to pass through freely. The great interest in this developing research area is in deciding on the nature of the emitting objects within or beyond the dust clouds. It is obvious that a dust filter will turn any star into an infrared star, and it is necessary to separate out "infrared sources" which are really highly reddened ordinary stars from the regions genuinely radiating in the infrared. These dusty regions

Three telescope domes crown the 4200-metre summit of Mauna Kea in Hawaii. The dome in the foreground houses the 3.8-metre diameter UK Infrared Telescope (UKIRT), the world's largest infrared instrument. The other domes contain the University of Hawaii's 2.2-metre telescope (centre), and the 3.6-metre Canada–France–Hawaii telescope (background)

indicate stars in the process of formation, sometimes singly, often in associations. The exciting prize lies in being able to study stars, and possibly new planetary systems, in the process of condensation.

The new telescope takes full advantage of the two main "windows" in the atmosphere which allow infrared to pass down to low altitudes, particularly the 10-micrometre band. (The other is close to 2 micrometres.) Astronomers also want the best possible resolution, and the original specification of 2–3 arcseconds resolution was upgraded in 1977 to 1 arcsecond, for a modest £12 000 on a total of £2½ million budget. The money appears to have been well spent, for the final performance is slightly better; finer than the resolution of most optical telescopes.

Normal telescope mirrors require thick slabs of low-expansion glass or similar materials such as Zerodur so that the optical figure does not change due to flexure or temperature variations. Suitable blanks are in short supply. They are very expensive and lead to a number of complications; not least the sheer time needed to cool and anneal them before any optical work can begin. The ratio of diameter to thickness for conventional mirrors is about 6, which, for big telescopes, means a mass of many tonnes and a correspondingly massive telescope support. The UKIRT mirror has a ratio of about 15, on a diameter of 3.8 metres. With such a thin and lightweight mirror it has been possible to reduce the weight, and hence the cost, of the remaining telescope structure. The latter, including the mirror-location cell, has been manufactured in Sheffield by the steel firm of Dunford Hadfields Ltd. The result is a telescope of unique design which has cost only a fraction of a comparable optical counterpart.

A television guidance system with computer control can use stars as faint as 5th magnitude under daylight conditions. The computer control can also make effective allowances for any structure flexure to keep the pointing accuracy within very fine tolerances. This circumvents the main problem with infrared detection systems, the difficulty of finding very weak sources against the powerful background radiation from the sky and the telescope itself. The problem has been likened to seeing a glowing cigarette end in a burning haystack!

Spurious signals generated by the warmth of the detector itself can be a problem since the incident energy is only 10^{-11} watts for the faintest detectable objects. Hence cooling to cryogenic temperatures is necessary. A full range of cryogenic handling facilities is being provided at Mauna Kea for all possible detectors. As with terrestrial infrared spectrometry, the background for these faint sources is cancelled by introducing an alternating signal (in this case by vibrating the secondary mirror) which can be amplified by an electronic AC amplifier. Care has to be taken to avoid microphonic noise, caused by the vibration, and to eliminate unwanted radiation from the relatively "hot" parts of the telescope in the optical path.

Cosmic dust particles are often of a roughly oblong shape, and will align themselves parallel to magnetic fields in space. Consequently, a study of the polarisation of infrared light can yield a great deal of information on the dust density or the magnetic field in the simple interstellar dust clouds, or on the magnetic forces acting on the star-forming regions. The role of magnetic fields in the formation of stars and planetary systems is incompletely understood. To gain full benefit from a programme on polarisation, beyond that already envisaged, astronomers will need to compile a catalogue of standard polarisation stars in the infrared, related to the parameters already known at other wavelengths. Identification of such reference stars will allow a more detailed study of dust clouds in depth, or of star formation regions in the line of sight of the standard.

Besides infrared work, the new telescope should also be ideal for submillimetre astronomy, through the three atmospheric windows at 1.2 millimetres, 800 micrometres and 400 micrometres. The 1.2-millimetre band is particularly important because molecules with two heavy atoms are detected in this spectral region. It will allow a study of cosmologically significant isotopes and molecules, and the overall chemistry in interstellar regions. Molecular lines of carbon- and nitrogen-containing molecules such as CO, $C^{18}O$, ^{13}CO and HCN occur in this band. A natural link with infrared investigations of the star condensation regions is provided here with CaH radiating in the same band. This is a common component of cosmic dust and can be related to silicates, a major component detected around 10 micrometres in the infrared. MgH and AlH^+ are detected in the 800-micrometre submillimetre band. Further research areas for such a combined or separate study could include other molecules or kinematics in galaxies, particularly near galactic centres.

One problem associated with the new telescope is the medical hazard of working at 4200 metres (14 000 feet), particularly in regular commuting to these altitudes. An acclimatisation period of at least 24 hours at about 2800 metres is necessary and a mid-level camp exists for this purpose. Full use of the 40-centimetre autoguider telescope, visual display units and other computer-automated controls should reduce the inefficiency that astronomers are likely to experience because of the altitude. Experience with the Anglo-Australian Telescope, operating at a lower altitude, has shown that it pays to have as many astronomers as possible on site handling the data to gain the maximum benefit available from the observing time. A simulator has been built at Edinburgh so that potential users can check in advance that their apparatus will plug in when it gets to Mauna Kea. It is unreasonable to expect an observer to waste time on site and simultaneously contend with altitude sickness.

22

Controlling UKIRT by trunk call

"THIS WEEK"
16 September 1982

The British infrared instrument is the first ground-based telescope to be operated by intercontinental remote control – with messages carried by ordinary telephone lines.

Astronomers at the Royal Observatory, Edinburgh, have successfully tested a link between their Scottish headquarters and the UK Infrared Telescope on Hawaii. This is the first time that astronomers on one continent have directly controlled a telescope on another, and it is a major step towards the remote operation of all Britain's major telescopes at good observing sites abroad from control rooms in Britain, thus saving travel costs and making the operation of the telescopes more flexible.

The telescope on Hawaii is the biggest single-mirror infrared telescope in the world and is on the best site for this kind of observation: the 4200-metre summit of Mauna Kea in Hawaii, above the atmospheric water vapour and carbon dioxide that absorb infrared radiation from space. Astronomers from the Edinburgh Observatory, which operates the telescope, had already installed a remote-control room at sea level in Hawaii to avoid altitude sickness. The control room is linked to the telescope by standard telephone lines, which provide television pictures of the image the telescope "sees" as well as main observational data such as the infrared spectra of stars.

This control room is now linked to Edinburgh. The new link carries less information but is adequate for messages controlling the telescope's motion and displays of the spectra of stars. It also allows staff at either end to type messages to each other to coordinate the telescope's activities.

The link was not simple. A console in Edinburgh passed messages

Observing by remote control: astronomers at the Royal Observatory, Edinburgh, Malcolm Stewart and Tim Hawarden, operate UKIRT from a distance of over 1100 kilometres. On the screen is an infrared spectrum of a star secured by the telescope a few seconds previously

via the observatory's computer to the Edinburgh Regional Computer Centre. This is connected, via the Science and Engineering Research Council's Sercnet computer network, to the Rutherford Appleton Laboratory in Oxfordshire. From there, messages pass to British Telecom's Packet Switching Service, then to the International Packet Switching Service and by satellite to the USA's packet switching network called Telenet. The message emerged at the main Telenet node for Hawaii in Honolulu and passed over leased telephone lines to the telescope's control room at the foot of Mauna Kea.

The three-hour test cost only £80. The complexity of the link means that there is a gap of several seconds between the sending of a message from Edinburgh and its arrival in Hawaii, although the link will be simpler in future. The astronomers plan to maintain the link with Sercnet because this also connects astronomers in five other centres through the Starlink system.

23

Infrared satellite blazes the trail

CHRISTINE SUTTON

27 January 1983

The first infrared satellite, IRAS, found more sources *each day* than the total listed in previous catalogues. This astonishingly successful satellite has probably pinpointed a million sources in all, for investigation by ground-based telescopes and future infrared observatories in orbit.

IRAS, for Infrared Astronomy Satellite, is a project developed jointly by scientists and engineers from the Netherlands, the United States and Britain. Its purpose is to survey systematically the whole sky in the infrared, from a vantage point well above the obscuring effects of the Earth's atmosphere.

Launched by a two-stage Delta rocket from NASA's Western Test Range, off Vandenberg Air Force Base, California, on 26 January, 1983, IRAS took nearly 10 years to progress from the drawing board to lift-off. The satellite was first conceived as a successor to the ANS (Astronomical Netherlands Satellite) which the Netherlands Agency for Aerospace Programmes (NIVR) had flown in the early 1970s. While the ANS had been successful in pioneering studies of the ultraviolet and X-ray sky, the new satellite would turn to the longer wavelengths of the infrared. NASA and the British Science Research Council soon became interested in the prospective infrared mission, and in October 1977 the three countries signed contracts formalising the collaboration.

The course was not all plain sailing, as changes in the launch date, originally set for August 1981, reflect. The problems that have beset the project, however, are a result not of misjudgement or incompetence but of real difficulties associated with infrared astronomy, which have meant that IRAS has required certain aspects of space technology to be developed well beyond previous limits.

The primary problem that all infrared astronomers face, whether using instruments on Earth or on a satellite, is that all objects emit infrared or "heat" radiation. Thus a telescope designed to detect infrared radiation itself emits some radiation as a "background", from which the image (or "signal") that the telescope is supposed to detect must be extracted; statistical fluctuations in the radiation, known as "photonic noise", cause further difficulties. The telescope for IRAS, built in the United States, overcomes these basic problems by carrying with it its own giant cooling system, around which have centred some of the problems that have delayed IRAS.

The satellite's telescope is surrounded by a huge double-walled dewar, a vessel like a Thermos flask, filled with liquid helium. The liquid cools the telescope's mirrors to around 16 K, thereby minimising the thermal background from the telescope itself. The dewar, 3.6 metres high and 2 metres in diameter, holds 70 kilogrammes of liquid helium, sufficient to keep the telescope cool for several months, until the helium has boiled away and the temperature of the whole assembly rises rapidly. The dewar is the largest ever to have been flown in space, although smaller devices have operated successfully.

The telescope on IRAS is a modified form of the so-called Cassegrain design, employing two mirrors. A concave primary mirror, made of beryllium and 57 centimetres in diameter, reflects incoming light onto a small convex secondary mirror situated 75 centimetres above the primary. The secondary mirror focuses the light through a small hole in the primary mirror, down onto an array of 62 detectors located in the telescope's focal plane. The detectors are strips of silicon and germanium, "doped" with small amounts of arsenic and gallium respectively. These semiconductors have been selected to respond to infrared radiation in four different wavebands, around 12, 25, 60 and 100 micrometres, a range of wavelengths that covers the maximum emission from bodies radiating at temperatures from 400 down to 40 K. The array is used to pick up infrared sources within the telescope's field of view of ½ degree, that is, about the size of the Sun's diameter as seen from Earth. A knowledge of the telescope's direction then gives the positions of the sources typically to within ½ arcminute.

Also situated in the telescope's focal plane are three instruments that scientists from the University of Groningen in the Netherlands have built. One additional device is a spectrometer to study the infrared spectra of strong point sources. Another is the so-called "chopped photometric channel", which has a smaller field of view

The Dutch–American–British infrared satellite IRAS being made ready for its launch early in 1983. IRAS was controlled from a specially built centre at the Rutherford Appleton Laboratory in Oxfordshire

than the main survey instrument. It maps the absolute intensities of individual objects, such as galaxies and dust clouds, at two wavelengths, 60 and 100 micrometres. Finally, the Dutch scientists have provided a short-wavelength device (5 micrometres) with a small field of view which will essentially count stars.

The Netherlands space agency, NIVR, has been responsible for the spacecraft that carries the telescope, and which houses the instruments to control the satellite's "attitude", or pointing direction, and to store the data collected. The satellite carries a tape recorder which can hold up to 14 hours' worth of data. This disgorges the stored information twice a day as IRAS passes over the operational control centre at Chilton in Oxfordshire. The antenna that picks up the signals is a 12-metre dish built in the United States and specially erected at the Rutherford Appleton Laboratory at Chilton.

Once received, the data is converted from radio to microwave signals and passed along cables to the control centre nearby. A team of some 100 scientists and engineers at the laboratory, including about 15 astronomers from all three participating countries, performs a quick, preliminary computer analysis of the data. They not only gain a first impression of the infrared sky but also check that the satellite is functioning properly. The team can, for example, check the efficiency of the semiconducting detectors by comparing the new data with that for known infrared sources. Such information allows them to plan ahead, using data from the previous period perhaps to make corrections after the next pass of the satellite over Chilton.

The antenna also transmits to IRAS during the short interval the satellite is overhead. The transmitted message contains details of all that the satellite has to do in the next period, in particular to implement the survey. It may also include instructions to observe particularly unusual sources; to repeat collection of data that have been lost or spoilt in some way; or to take corrective actions based on data already received.

IRAS's orbit takes it round the Earth every 100 minutes, passing within 9 degrees of the poles on a path 900 kilometres high, well above the water vapour in Earth's atmosphere which absorbs much of the infrared radiation from space. During one orbit the satellite observes a band of sky half a degree wide. The telescope can scan two opposite segments of sky, each 30 degrees wide, in 15 days. During this time the field of view scanned on successive orbits overlaps by ¼ degree, ensuring that suspected infrared sources can be sufficiently well confirmed to support the astronomers' aim that the survey will be 99.8 per cent reliable. The precession of the satellite's orbit – the way it swings slowly round the Earth – means that successive 30-degree segments overlap by 15 degrees, so assisting the procedure of confirming sources.

At times when the satellite is not involved in surveying, for example when passing near the poles, IRAS observes objects that are particularly interesting astronomically. In this mode the orbiting telescope operates like one on Earth, following suggestions put forward by scientists from all three participating countries.

What do astronomers hope ultimately to learn from IRAS? The

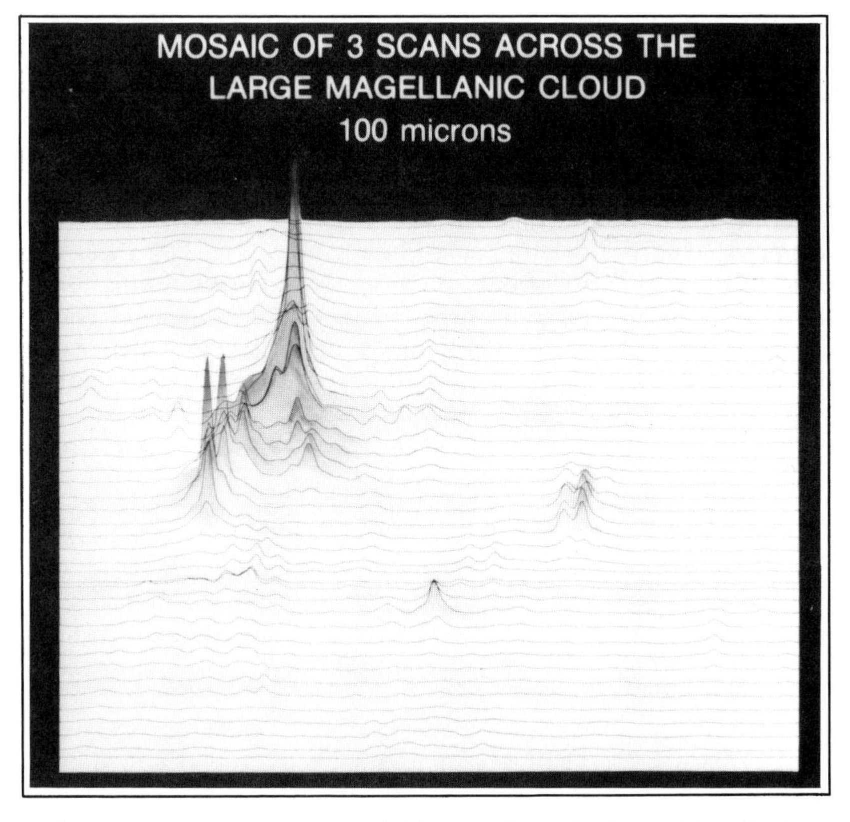

Scans across our nearest neighbour galaxy, the Large Magellanic Cloud, by IRAS, show strong emission at 100-micrometre wavelength from the region of the Tarantula nebula (left of centre). The radiation comes from vast clouds of warm dust surrounding this region of active star formation

satellite's primary role is to fill in the existing picture of the heavens, revealing stars, galaxies, clouds and so on that are not seen at the shorter optical and "near" infrared wavelengths (less than 5 micrometres), or at the longer wavelengths observed by radio telescopes. Objects may appear at infrared wavelengths because they are cool, because they are obscured by dust that transmits only longer wavelengths, or because their intrinsic mechanism of radiation cuts off at wavelengths shorter than infrared. So far, infrared studies at observatories such as the UK Infrared Telescope (UKIRT) on Hawaii have centred only on those objects astronomers already know about from information at optical and radio wavelengths. IRAS's survey will, however, be totally unbiased by existing knowledge, and should thus provide a completely revised picture of the infrared sky. While discussions of what astronomers expect to find can only be expressed in terms of what they already know, those working on IRAS clearly hope that the satellite will make exciting new discoveries.

The survey data are being analysed fully at the Jet Propulsion Laboratory in California, where a catalogue of all sources discovered, expected to number in the region of 1 million, will ultimately be made. This should provide enough new information to keep astronomers busy for many years to come, those in Britain being particularly well placed to follow up the survey with work on UKIRT on Hawaii. Clearly infrared observations are set to make a large impact on astronomy and astrophysics during the next few years.

24

First results from IRAS

"MONITOR"

2 June and 10 November 1983

"THIS WEEK"

18 August and 10 November 1983

The IRAS cooling system worked better than expected, extending the satellite's life from a predicted six months to almost a year. Both its survey and the detailed observations turned up surprises, even during the first few months of observation, although a full analysis of its data will take years.

Hot on the heels of the Japanese discovery of a ring close to the Sun, comes the IRAS observation of what appears to be three more rings of dust circling the Solar System. Their temperature (around 150–200 K) suggests that they lie near the asteroid belt, between the orbits of Mars and Jupiter. The rings are probably the results of asteroid collisions. The broad central band could have built up slowly during millions of years of the steady, microscopic chipping of boulders bumping around in space.

The other two bands may be the product of a single cataclysmic event, perhaps caused by a comet flying through the asteroids towards the Sun and demolishing an asteroid. Eventually, the dusty remains would have spread out to circle the Sun completely. The dust particles continue to orbit slightly irregularly, spending more time at the extremes of their wobble and thus appearing as two distinct bands.

IRAS has also found an asteroid-like body which can approach the Sun to less than half the distance of Mercury. The object, catalogued 1983 TB, may turn out to be a "missing link" between asteroids and comets.

1983 TB was identified on the IRAS Sky Survey by Simon Green of Leicester University. The Leicester team has a project to pick up fast-moving objects in the IRAS data, with the aim of identifying asteroids with unusual orbits which cross that of the Earth. Until now the project has borne more exotic and unexpected fruit. Instead of asteroids, the team has kept on discovering new comets, of which the most famous (out of a total of five) was last spring's IRAS-Araki-Alcock.

1983 TB has turned out to be more exotic than expected. Its 1½ year-long orbit makes it an Apollo asteroid, one of a group of some 50 small bodies whose orbits cross the Earth's. But it stands out from the rest by swinging closer to the Sun than any other asteroid known. Its closest approach to the Sun is only 20.7 million kilometres – beating the previous record-holder, Icarus, by 5 million km.

More unusually, the orbit of the newly discovered asteroid exactly matches that of the stream of debris which gives rise to the Geminid meteor shower, visible between 12 and 15 December each year. Meteor showers are caused by small particles of dust from decaying comets entering Earth's atmosphere, and burning up as "shooting stars". All the main meteor streams – such as the Perseids, the Lyrids and the Leonids – have associated "parent" comets which shares the same orbit. However, the Geminids, which can produce up to 60 meteors per hour, have – until now – had no known parent body.

In fact, 1983 TB may not be an asteroid after all, but a severely decayed comet. There have long been suspicions that Apollo asteroids are extinct comets. Repeated passages of a comet close to the Sun can drive off all its volatile ices, which form a large gaseous head around the nucleus of an active comet. But when these have dissipated, the very small nucleus would look similar to an asteroid.

As well as its discoveries within our Solar System, IRAS has provided the strongest evidence that our planetary system is not unique. Its discovery of infrared radiation in the vicinity of Vega – the brightest star directly overhead in summer – is the first direct detection of planetary material around another star.

IRAS was checking on bright "calibration stars", one of which was Vega. IRAS normally investigates "lukewarm" bodies but there is little about Vega that is lukewarm. It is twice as massive as the Sun, has a surface temperature of more than 10 000 K, and is 58 times more luminous than our star. So a team headed by Americans George Aumann and Fred Gillett, were surprised to find that Vega has a powerful excess of infrared radiation, particularly at the longer wavelengths which come from cooler matter.

Aumann and Gillett repeatedly cross-checked their data in case it could be due to an infrared source behind or in front of the star, but as Gillett eventually concluded, "we're convinced it's coming from the immediate vicinity of Vega". After rough calculations, the two astronomers processed the data fully at the Jet Propulsion Laboratory (JPL) in Pasadena, which is compiling the IRAS catalogue. Aumann says, "The search was real detective work. You have the fingerprints, you have the weapon . . . then you go all-out to find the culprit."

The "culprit" seems to be a disc of dust particles, each about a millimetre in size and at a temperature of 90 K. The disc stretches to nearly twice the extent of our Solar System. Aumann points out that we are probably seeing the main mass of Vega's planetary system, which would form further away from its luminous "sun" than the family of planets around our Sun. IRAS cannot detect the radiation from any of Vega's planets directly, because these concentrations of matter would have relatively less surface area than the dust.

But if Vega's system contains about the same proportion of small to large bodies as the Solar System, the astronomers calculate that the mass of Vega's disc is around one-thousandth that of the Sun – very similar, in fact, to the mass of our Solar System. Gillett cautiously admits that this is "an interesting figure".

If there are planets around Vega, what state are they in? Vega cannot be more than a few hundred million years old, so its planetary system must be at a relatively early stage of its evolution. Planets have probably formed, but with a great deal of dust and debris still present. Our Solar System is 4600 million years old. Life took hundreds of million of years to develop on the Earth, so even if Vega has a suitable planet, life has almost certainly not evolved yet, and Vega is the type of star that is short-lived and so will "die" before life could get under way properly.

Vega is not a typical infrared source. Most astronomical sources of infrared radiation are clouds of dust in space, warmed to temperatures between about 30 K are a few hundred K. The densest clouds are apparent to optical astronomers only as dark silhouettes, blocking off the light from stars beyond. But new stars are forming from the agglomeration of gas and dust at the cloud's centre. As the material of these protostars condenses, it becomes warmer and emits infrared radiation. This can penetrate the surrounding dust cloud, and can be picked up by infrared telescopes.

Ground-based infrared telescopes have detected radiation from massive and powerful young stars, some 20 times heavier and 10 000 times as luminous as the Sun. These lie within very large, rare clouds. Sun-like stars are thought to form in smaller, more common clouds. They emit less infrared, and have so far been undetectable. IRAS has now found such low-mass protostars for the first time, lying in two dark clouds called Barnard 5 and Lynds 1642 (after the astronomers who first catalogued them). The latter contains one or two protostars, while the former has a central cluster of five protostars. These embryonic stars are less than one million years old (very young in astronomical terms), and further investigation should show whether they have surrounding discs of dust which are condensing into planets, like those of the Solar System or of Vega's suspected system.

IRAS is also investigating star formation in other galaxies. Its views of spiral galaxies show up the spiral arms – but these infrared-bright regions do not correspond exactly with the concentrations of stars seen in the optical photograph. They coincide instead with the

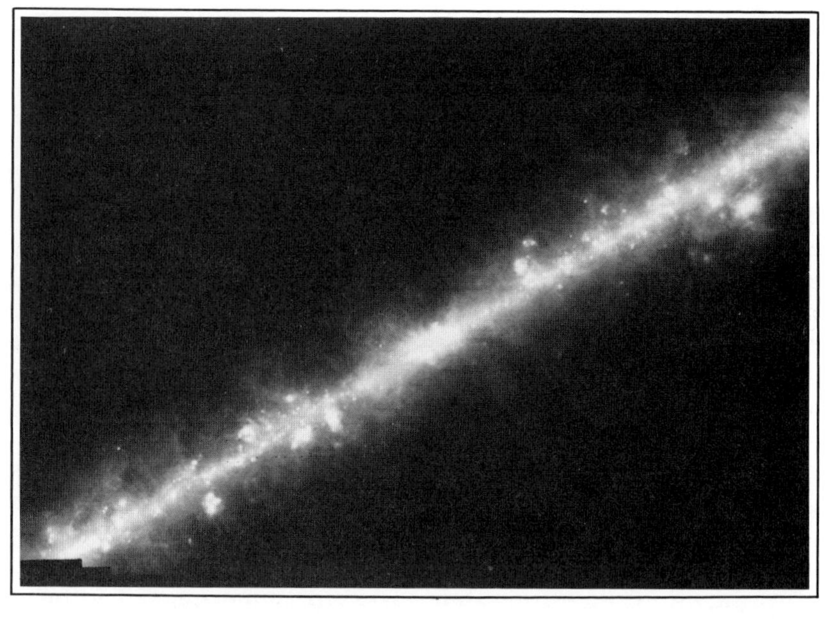

IRAS's view towards the centre of our Galaxy (middle of picture) penetrates the dust that obscures the optical view. The band of the Milky Way stretches diagonally across the infrared view, here processed as a photograph. The knots and blobs along the band are giant clouds of gas and dust where new stars are being born, while above and below the band lie faint wisps of the newly discovered "infrared cirrus".

narrow dust lanes seen in silhouette at optical wavelengths. The infrared picture therefore actually gives us a preview of the galaxy's next generation of stars.

In spiral galaxies, the existing gravity of the star has compressed the galaxy's gas and dust to form new stars along the spiral arms, as predicted by current theories of spiral structure and star formation. But IRAS has found unpredicted regions of star formation in other galaxies. A pair of galaxies passing close to one another, NGC 1888 and 1889, are surprisingly strong infrared sources. Astronomers have known for years that their gravitational interaction is pulling off stars from their outer regions, but these new observations show that the interaction is also somehow compressing interstellar gas and dust within the galaxies, to form new stars.

As well as the star-formation regions, IRAS is revealing that much of interstellar space in our Galaxy is littered with wispy clouds of dust and gas, known as infrared cirrus. Their brightness around the infrared wavelength of 100 microns suggests that they are made up of graphite dust that has been ejected from stars as stellar wind and mixed up with ionised hydrogen gas. Particularly striking are wisps of material that apparently come from the centre of the Galaxy. Only in the infrared are these structures revealed with such clarity.

The satellite has detected more than 200 000 infrared sources so far – 100 times more than were previously known. Many do not fit into existing astronomical categories. Some 130 cold, point-like sources have been selected for further investigation from an initial list of 8709 objects. Nine of these sources are completely new to science; it is not even known yet how far away they are. They could be just beyond the Solar System or, more likely, represent distant very young galaxies packed with hot, bright stars.

25

A more penetrating infrared eye in space

NIGEL HENBEST

31 March 1983

The European Space Agency's next major project will be a £150 million Infrared Space Observatory (ISO) to follow up the pioneering survey by the IRAS satellite.

A superficial description of ISO does not sound very different from IRAS. Its telescope will have a mirror 60 centimetres across, and will be cooled by liquid helium to a temperature of only a few degrees above absolute zero; and its detectors will be sensitive to wavelengths up to a maximum of 120 micrometres. But IRAS is only a survey telescope, with simple detectors and a planned lifetime of around seven months, during which its main task is to sweep the entire sky and catalogue the sources it finds. ISO, on the other hand, is an observatory. Like observatories on the ground, or the orbiting International Ultraviolet Explorer, ISO will be pointed to particularly interesting astronomical objects for thorough study. To achieve this, ISO must have a longer lifetime than IRAS.

Infrared telescopes must be cooled to within a few degrees of absolute zero to prevent their own heat radiation from swamping the faint signals from space: hence the liquid helium cooling systems which make infrared satellites complex and expensive. Once the coolant has all evaporated, the telescope warms up and is useless. ISO uses 100 kilogrammes of super fluid liquid helium at 3 K to cool the detectors and the telescope, but – unlike IRAS – it also carries 50 kilogrammes of slightly warmer liquid hydrogen to provide the bulk of the cooling for the satellite. The lifetime of ISO should be at least 1½ years, and judging by the success of the IRAS system it may well be 2½ years – as long, for example, as the Einstein X-ray Observatory survived. ISO will follow an elliptical orbit taking it from 1000 to 39 000 kilometres above the Earth's surface, and will complete one orbit every 12 hours. Instead of recording data, it will transmit them back continuously to tracking stations at Villafranca in Spain and Carnarvon in Australia. ESA hopes to launch the 1.8-tonne ISO on an Ariane rocket in about 1990.

The sources that ISO will be investigating will generally be clouds of dust in space, with temperatures from a few hundred degrees down to only 30 K. The warmer dust clouds are found close to stars, either in dark nebulae adjacent to stars that have just formed, or in the ejected shells of matter from old stars, called planetary nebulae. Long-wavelength infrared comes from cooler clouds that are just beginning to contract to form stars. ISO should reveal for the first time the fine details of star birth.

Many galaxies beyond our own also emit a large proportion of their radiation – sometimes most of it – as infrared. It can come from dust between the stars, but distant quasars and other active galaxies produce powerful infrared emission from their cores, which probably contain gas swirling round a massive black hole. Looking beyond these, and hence farther back in time, ISO may "see" the birth of galaxies soon after the big bang, something ground-based infrared astronomers have sought for years.

ISO has three detector systems for astronomers to position at the telescope's focus to investigate different aspects of infrared sources. A large semiconductor (indium antimonide) chip divided into 1024 radiation-sensitive segments can "photograph" the source at the shorter wavelengths of 1–5 micrometres. It may be paired with a newly developed silicon-bismuth chip which will simultaneously "photograph" at 5–18 micrometres.

The second system is a photometer to measure the brightness of a source simultaneously at 12, 40 and 100 micrometres with three chips made of silicon–gallium, germanium–beryllium and germanium–gallium respectively. The photometer also incorporates subdivided chips to produce crude pictures at these wavelengths. Wire grids can be inserted in front of the photometer to show the polarisation produced when infrared radiation is reflected from dust grains. Finally, a pair of spectrometers (Michelson interferometers) can split the radiation up into detailed spectra, covering the range 2–70 micrometres. From these spectra, astronomers will be able to measure the conditions in gas clouds that are obscured at optical wavelengths by the ubiquitous dust – for example, in regions where stars form, and at the centre of our Galaxy. The spectrometer can also show absorption at specific wavelengths by the dust itself, thus helping to unravel the mystery of the dust's composition.

26

The race to explore the infrared-radio borderland

NIGEL HENBEST

30 June 1983

The wavelengths that border infrared and radio have produced exciting new results, especially in the study of star-formation. Several countries are now racing to build telescopes to observe these millimetre and submillimetre waves.

In April 1983, bulldozers began to level a patch of ground near the summit of a mountain in Hawaii, to accommodate Britain's newest telescope – the world's largest instrument for exploring the heavens at the little-studied wavelengths between infrared and radio. Built in collaboration with the Dutch, the precision-built dish, 15 metres across, promises to reveal details of the birth of stars, as well as studying celestial objects at distances ranging from comets to the most distant quasars.

The telescope will pick up submillimetre waves, radiation with wavelengths between 0.3 and 1 millimetre. The study of these waves, and the slightly longer millimetre waves (1–10 millimetres) is a new branch of astronomy, untouched until the 1960s by either the infrared astronomers studying shorter waves, or radio astronomers working longwards of 10 millimetres. These two branches of astronomy are divided by a natural barrier: the atmosphere's absorption of radiation from space. Infrared astronomers began by working to longer wavelengths from the red end of the visible spectrum, investigating the sky at the few wavelength bands ("windows") where the atmosphere does not absorb too strongly. Between 40 and 300 micrometres (0.3 millimetres), the atmosphere is completely opaque. At the other side of this divide, the submillimetre waves can penetrate the air to a certain extent. They are absorbed by water vapour, but they can be detected from a suitably high and dry site. But to most infrared astronomers, these waves were on the radio astronomers' side of the divide.

The breakthrough came with the development of high-precision dishes. John Findlay, of the American National Radio Astronomy Observatory, constructed a medium-sized radio telescope, 11 metres in diameter, to focus radiation as short as 2 millimetres. To avoid much of the water vapour, this 11-metre diameter dish was erected on top of 2100-metre high Kitt Peak in Arizona in 1967, alongside the optical telescopes of the separate Kitt Peak National Observatory. It was equipped with an infrared detector invented by pioneer infrared astronomer Frank Low; unlike other infrared detectors, it was sensitive to far infrared and millimetre waves as well as near infrared.

With the new 11 metre dish, Low made the first studies of the brightness of sources at millimetre wavelengths. Such measurements have grown to become one important facet of present-day submillimetre- and millimetre-wave research; accounting for perhaps 20 per cent of the time spent on modern telescopes operating at these wavelengths. The observations reveal the coolest of interstellar dust clouds, with temperatures only a few degrees above absolute zero. They also allow astronomers to investigate the outbursts in the distant exploding galaxies called quasars, which should occur sooner at these wavelengths than at the longer wavelengths detected by traditional radio telescopes.

But the bulk of modern research at the submillimetre and millimetre wavelengths – and the most exciting results – is in the field of studying the spectra of gas in space. This branch of the subject started from the application of radio techniques. In the late 1960s, communications engineers at Texas University and the University of California at Berkeley designed and built relatively small (5- and 6-metre diameter) radio dishes for millimetre waves, detecting the radiation with newly designed, very-high frequency radio receivers. By tuning the receiver, an astronomer could investigate in detail the spectrum of a source, as well as its total brightness.

But what could produce a spectrum at these wavelengths? They correspond to very small changes in energy, and the only possible candidate was the slight alteration in energy as a spinning molecule changes its rate of rotation. In the 1960s, this did not seem a fruitful field. Although optical astronomers had found signs of three simple, two-atom, molecules in 1937–41 and radio astronomers in 1963 had picked up 18-centimetre radio waves from hydroxyl molecules (OH), most astronomers believed that molecules would be very rare

in space. It was unlikely that more than two atoms would meet up and combine, and the fragile molecules would anyway be broken up by ultraviolet radiation from hot stars.

American physicist Charles Townes did not accept this orthodoxy. In 1953, he had produced the first maser, using ammonia molecules, and in 1968 he used the 6-metre dish to look for radiation from ammonia in space. After much effort, his team detected its characteristic emission at 13 millimetres, and very soon afterwards strong emission from water molecules at an adjacent wavelength. That year marked the true birth of millimetre-wave astronomy. If three- and four-atom molecules (water and ammonia) could exist in space, there must be others. Although most molecules produce spectral lines at wavelengths of several centimetres, which can be detected by conventional radio telescopes, the strongest emission is at the submillimetre and millimetre wavelengths.

The discovery of interstellar molecules opened up a whole new field of science: astrochemistry. It is concerned with reactions between atoms and molecules in conditions of extremely low density, pressure and temperature, which cannot be achieved in the laboratory. Astrochemistry is a slow business, where the molecules are built up one atom at a time, with the chain started when a cosmic-ray particle happens to ionise an atom. The reactions occur in dark, dense clouds, where the molecules are protected from ultraviolet radiation. Study of different molecules, and of various spectral lines due to the same molecule, can reveal details of the density and temperature throughout the cloud, while the Doppler shifts of the lines show the motions of gas in the cloud – all hidden from optical astronomers by the dust.

One of the major astronomical discoveries of the 1970s came from the study of carbon monoxide in space. The molecule occurs in huge molecular clouds found mainly in the inner region of the Galaxy. The lines from carbon monoxide indicate that the density of these clouds is far higher than expected from the molecules alone, or from the density of hydrogen atoms there, as revealed by their well-known emission at a wavelength of 21 centimetres. The extra

---

(Opposite) *The world's first telescope specially designed to detect radiation at millimetre wavelengths, between infrared and the radio band. Its 11-metre dish is protected from the Sun and wind by a dome. The instrument is at a height of 2000 metres on Arizona's Kitt Peak, above most of the atmospheric water vapour which absorbs radiation of these wavelengths*

gas can only be in the form of hydrogen molecules, which we cannot observe directly. Allowing for the number of molecular clouds, and their size and density, astronomers calculate that there must be as much molecular hydrogen in our Galaxy as there is atomic hydrogen – effectively doubling the known amount of gas in the Milky Way!

At the centres of the clouds, the concentrated gas and dust are forming into stars. Studies of molecular lines show that the gas and dust do not just collapse as shrinking spheres. They form into dense rotating discs, which eject more tenuous gas at high speed from either side. These wavelengths are providing the best information we have on the long-debated topic of how stars and planets – including the Solar System – came into being.

In the 1970s, several more millimetre wave telescopes were built, in New Jersey, Massachusetts and Sweden. Other countries are now wanting to get into the field of millimetre waves, and the shorter submillimetre waves which have been scarcely touched. There are three approaches. Build a big telescope to work at millimetre wavelengths with greater sensitivity than before; or erect an array of telescopes to achieve very high resolution of details; or build a smaller but highly accurate dish to focus the submillimetre waves. The Japanese have taken the lead in the millimetre-wave field – at least the longer end of it. In 1982, they completed a radio telescope 45 metres in diameter (over half the size of the dish at Jodrell Bank) yet so accurately made that it can focus radio waves as short as 3 millimetres. In terms of surface smoothness relative to its diameter, this Japanese dish is the world's best radio telescope. Next to it, they are constructing an array of five 10-metre telescopes which can "see" finer details; its resolution is equivalent to that of a single dish half a kilometre across. The Japanese government paid the equivalent of £25 million for the whole installation, which was designed and constructed by groups from Japanese universities and Tokyo Observatory working with companies like Mitsubishi.

There is also to be a large telescope and an array in Europe, operated by the German–French Institute de Radio-Astronomie Millimetrique (IRAM). The organisation has its headquarters in Grenoble, and it was originally to have had a 30-metre dish and an interferometer array on the same site. The Germans have constructed the big dish, and to get above the worst of the water vapour, they have sited it at an altitude of almost 3000 metres, at Pico Veleta, in the Sierra Nevada near Granada. German astronomers hope the 30-metre telescope will be able to take full advantage of its

Three of the five dishes of the interferometer at Japan's Nobeyama Radio Observatory. Each is 10 metres across and has a surface shaped to an accuracy better than 0.1 millimetres

site by working at wavelengths as short as 1 millimetre – and so, incidentally, overtaking the Japanese instrument as the most accurate radio telescope. The French are building the array. Since it requires a stretch of level ground, it has ended up in a different place from the German dish, at the Plateau de Bure, conveniently situated in the Alps just south of Grenoble. The Plateau de Bure is 2550 metres high, and is reached by cable car. The ambitious array of three 15-metre telescopes should be completed in 1986. At its shortest wavelength of 1 millimetre, the 250-metre long array should resolve details as small as one arcsecond, rivalling optical telescopes.

British, American and German astronomers are pioneering the complementary approach, and tackling the new territory of submillimetre-wave astronomy. Telescopes to work at these shorter wavelengths must have smoother surfaces still and can hence only

Model of the British–Dutch submillimetre wave telescope shows its construction from 276 individual panels to form an accurate dish 15 metres across. The telescope swivels about a horizontal axis, and rotates on a circular rail

be built to a smaller size. The engineering problems are immense; the dish must have a surface accurate to 1/30 millimetres, over a curved dish 10 or 15 metres across, and keep its shape as it is tilted. The German Max-Planck Institut für Radioastronomie is collaborating with the University of Arizona in a 10-metre submillimetre wave dish to be placed on the 3000-metre high Mount Lemmon in Arizona.

The shorter wavelength submillimetre waves, however, are best observed from higher sites – and the best existing observatory site for this research is the peak of Mauna Kea, in Hawaii, at an altitude of 4200 metres. This peak already has two large optical telescopes and two infrared telescopes, including the world's largest single-mirror infrared telescope, the UK Infrared Telescope (UKIRT). Astronomers from the California Institute of Technology (Caltech) have built a 10-metre submillimetre wave telescope which they plan to site just below the summit, in "Millimeter Valley", where it will be protected from the worst of Mauna Kea's high winds and from electrical interference.

Britain's submillimetre-wave telescope is the largest now under construction, with a diameter of 15 metres. The idea of a British millimetre or submillimetre wave dish has been in the air since about 1970, but has seen many vicissitudes in the planning stages. Several years passed in negotiations over British involvement in the international organisation IRAM, which eventually fell through largely

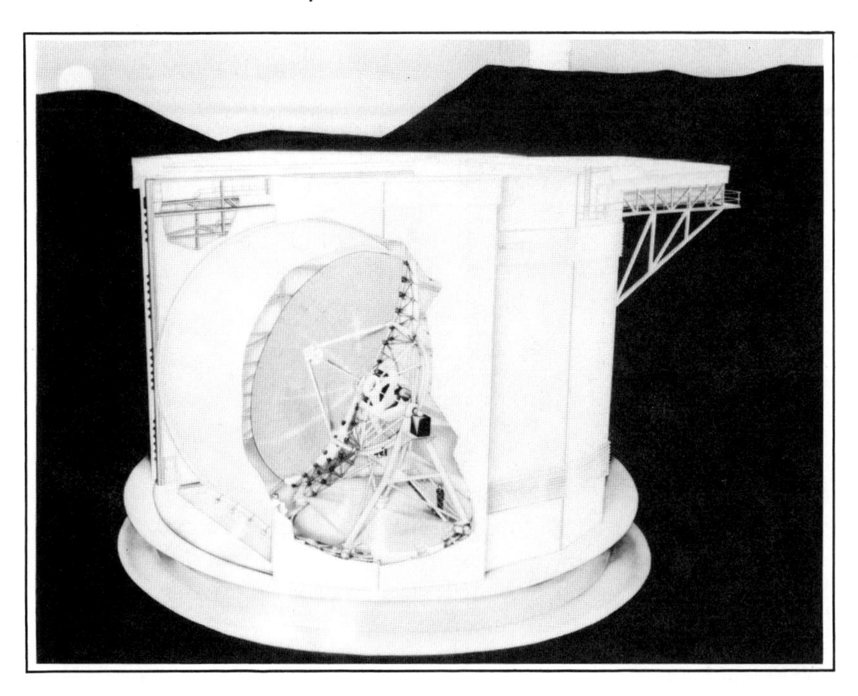

Artist's impression of the submillimetre wave telescope within its rotating building, in Millimeter Valley below the summit of Mauna Kea, whose existing telescope domes are shown. The building's shutters slide round to expose an opening (left) covered by thin PTFE cloth

because British scientists were not keen to join another international organisation with contributions in foreign currency. Pioneer millimetre-wave astronomers Richard Hills and Tom Phillips then put forward the proposal for the 15-metre British submillimetre wave telescope – to be the world's largest.

But there were problems with building the British telescope on the world's best site. The Hawaiian government was worried about the environmental effect of the telescopes already on Mauna Kea, and a proposed American 25-metre submillimetre-wave telescope (dropped in 1982) had priority over the British instrument. The British 15-metre telescope was officially approved in August 1980, when the UK had a suitable location, the site of the optical telescopes that the Royal Greenwich Observatory was building on the peak of La Palma in the Canary Islands.

The next twist in the saga came the following year, when Dutch astronomers became participants in the telescopes at La Palma, including the submillimetre-wave dish, contributing 20 per cent of the cost and effort in return for the same proportion of the observing time. They were keen to use the telescope at the shortest wavelengths that the atmosphere can transmit, in the window at 0.3–0.4 millimetres. These can only rarely be observed from the relatively low altitude of La Palma, 2400 metres, but from Hawaii, astronomers can use this window for more than a third of the time, when the atmosphere is extremely dry. Negotiations were reopened with the Hawaiians, and this time were successful. After a year spent preparing a report on the environmental impact of the telescope, its location on Hawaii was approved in the spring of 1983.

Although similar in outward appearance to existing radio dishes, the highly accurate surface required for submillimetre observations has led to many special features in the design. A single large metal dish several metres in diameter could not keep its shape accurately when it tips up to track sources and the gravitational pull on various parts of the dish changes in direction relative to the dish itself. Instead, the telescope relies on a metal framework especially designed to keep its shape. As the dish tips up, the edges sag by a couple of millimetres, which would normally affect the ability of the telescope to focus submillimetre waves. But the framework is designed so that despite the sag, its shape remains a paraboloid and so can focus radiation accurately. The sag simply moves the position of the focus by about a millimetre, and the receiver is moved automatically to compensate.

The telescope will be put together and tested in Britain, then taken apart and shipped to Hawaii where it will be put together again. If the panels can be lined up on the framework as accurately as expected, the surface's deviations from the required shape should be typically as small as 0.03 millimetres. This means the telescope will be able to work extremely well down to a wavelength of 0.6 millimetres, and will produce reasonable information down to the atmosphere's cut-off of 0.3 millimetres.

This accurate surface must be protected from strong winds, which would flex and even sandblast it; from snow and frost in winter; and from the deforming influence of the Sun's heat. It is thus to be housed in a cylindrical building a little larger than the telescope itself. The building has sliding doors to create a wide aperture for the telescope to look through, and it will rotate with the telescope as it tracks objects across the sky. To prevent wind and sunshine entering the building, the slit is covered over with PTFE (Teflon) cloth, which is transparent to submillimetre waves and is strong enough to cope with high winds.

The building accounts for more than half the total cost of the project, some £7 million. To erect such a building in the UK would cost about £1 million, but the problems of erecting it on top of a mountain half as high as Everest increases the cost several times over. The telescope itself will cost less than £2 million; while the other £1.5 million will be spent on the site on Hawaii and the receivers and the control systems.

The design of the telescope is being undertaken at the Science and Engineering Research Council's Rutherford Appleton Laboratory, which is also providing management for the project, under physicist Ron Newport. They have already started producing the panels, and will soon start placing contracts for the engineering of the telescope's framework. The computer systems and controls are being developed at Cambridge University's Cavendish Laboratory.

Receivers for submillimetre-wave telescopes are similar in principle to ordinary radio telescope receivers (except that no electronic device can amplify the very high frequency signal before further processing). The incoming radiation is "mixed" directly with a signal of roughly the same frequency generated in the receiver, and they "beat" together at a lower frequency, which can be handled and amplified with conventional electronic circuits. New mixers must be developed to deal with the frequencies involved in submillimetre work and the various groups involved are working on several types. The Rutherford Appleton Laboratory is producing a mixer for the longer submillimetre waves, which involves a solid-state device called a Schottky diode, already in use in millimetre-wave receivers. For the shorter wavelengths, Queen Mary College in London – long a leader in receivers for these wavelengths – is building a mixer based on indium antimonide which should work down to 0.4 millimetres. Dutch researchers at Dwingeloo are also working on mixers, and producing cooled amplifiers. Researchers at Caltech have pioneered a mixer which consists of a very thin slice of insulator sandwiched between two thin layers of superconducting metal and are using them as extremely sensitive detectors of millimetre waves. Researchers at the University of Kent and at Groningen in Holland are looking into ways of using such devices at submillimetre wavelengths, perhaps by incorporating different superconductors.

The British groups are well to the fore in developing receivers, largely because they have been using the best existing "submillimetre-wave" telescope on the ground: UKIRT on Hawaii. Although built primarily as an infrared telescope, UKIRT's 3.8 metre mirror is large enough to give reasonable resolution and sensitivity at submillimetre waves, and its surface, intended for shorter wavelengths, is certainly precise enough. When the 15-metre submillimetre-wave telescope is built, it and UKIRT will be operated as a common facility by UKIRT's controlling organisation, the Royal Observatory, Edinburgh. For much of the time, the telescopes will be operated by remote control, either from sea level on Hawaii, or from Edinburgh.

To some extent, the British dish and the neighbouring Caltech 10-metre telescope in Millimeter Valley will be in competition. The 15-metre instrument will be more sensitive at most wavelengths, because of its larger size, and will show finer details when observing at the same wavelength as the 10-metre. But the Caltech dish has the more accurate surface (errors typically 0.02 millimetre) and will produce the better results at the shortest wavelengths of 0.3 millimetres. One advantage of having the telescopes adjacent is that astronomers will be able to connect them as an *interferometer* and resolve extremely fine details at these wavelengths for the first time. Both telescopes should be completed about the same time. The British are ahead on constructing the building, while Caltech has a lead in having the telescope itself completed. A target date for both is set by the chance to observe Halley's Comet on its return to the inner Solar System, for it will be at its nearest and brightest in the winter of 1985–6.

The ultimate in submillimetre-wave astronomy would come from getting above the Earth's atmosphere altogether. Some of the most interesting results so far have come from a relatively small telescope, with a 0.9-metre mirror, which has the advantage of being flown in an aeroplane at a height of 12 500 metres. The water vapour in the atmosphere lies in a relatively low layer, and while Mauna Kea is above 85 per cent of the water vapour, the Kuiper Airborne Observatory flies its telescope above 99.5 per cent. Unique among submillimetre telescopes, it has virtually no atmospheric absorption to cope with, and it is the only telescope (apart from those flown especially on balloons) that can observe at the shorter wavelengths of the far infrared.

For the future, satellite observatories will undoubtedly "see" even better from space; the European Space Agency, the Americans and a joint Russian–French team have all proposed satellites to observe at far infrared and submillimetre wavelengths. But an aeroplane equipped with a 1-metre telescope would cost somewhere around £20 million, or three times the cost of the 15-metre ground-based telescope, while a satellite would be several times dearer still. While such high-flying telescopes may hold our hopes for the future, it is the present ground-based telescopes that will open our eyes to the Universe in this still-virgin region of the spectrum.

PART SIX

Exotic Messengers from the Cosmos

Most of our knowledge about the Universe has come from electromagnetic radiation, be it light or any of the other wavelengths from radio waves to gamma rays. But there are other messengers crossing the cosmos, and their arrival at the Earth can provide us with information we cannot find out in any other way – especially as they are generated in some of the most violent and energetic places in the Universe: the explosion of stars as supernovae and the centres of active galaxies.

The exotic messengers fall into two classes. One type, comprising only the cosmic rays, has been known for decades and was not predicted in advance. The other type consists of the many other kinds of particle and radiation that theory tells us should be reaching us from space, but which have yet to be detected.

Cosmic rays were discovered 70 years ago, when physicist Victor Hess took up an ionisation chamber in a balloon and discovered that it was recording more radiations at high altitudes. This radiation was clearly not coming from radioactive elements on the Earth, and so had to come from space. Although it was dubbed "cosmic rays", the culprit has turned out not to be electromagnetic radiation, but high-speed charged particles – protons and other atomic nuclei. Many of them are travelling faster than can be achieved in a particle accelerator on Earth, showing they must come from regions of exceptional astronomical violence. Medium-energy cosmic rays may be accelerated by pulsars, or more likely by the expanding shock wave from a supernova explosion blasting through the interstellar gas. But recent results indicate that the fastest cosmic rays come from beyond our Galaxy. They may be accelerated in the cores of nearby active galaxies, thought by many astronomers to contain a disc of hot gas spiralling down into a massive black hole.

Other fundamental particles should also reach us from space, but these are either hidden in the welter of cosmic rays, or reach us so rarely that none has yet been reliably detected, or interact so weakly with matter that it is difficult to build a suitable detector. Particles of antimatter, for example, do occasionally turn up amidst cosmic rays, but special techniques are required to sort them out from the others. Among the rare exotic particles that may have been detected just once are a tachyon (a particle travelling faster than the speed of light) and a magnetic monopole, a hypothetical particle with just a north (or a south) magnetic pole. It is difficult to refute a single claim, for an experimenter may have been lucky enough to observe just the one particle that hits the Earth every million years. But indirect evidence has led most scientists to doubt the claimed detections.

The neutrino, on the other hand, is a particle that certainly does exist, and neutrinos are undoubtedly pouring down from space all the time. But they interact so weakly with matter that most pass straight through detectors – and right through the Earth. The world's first neutrino telescope, a large tank of chlorine-rich cleaning fluid down a goldmine in South Dakota, has picked up neutrinos from the nuclear reactions at the Sun's heart, but the prospects for a neutrino telescope to detect events farther off in the cosmos still lie in the future. The events most likely to be detected are supernova explosions, for some theories predict that most of a supernova's energy is emitted as neutrinos.

Finally, there is the possibility of detecting another type of radiation – gravitational waves. Albert Einstein's general theory of relativity, the accepted theory of gravity, predicts that the acceleration of a massive body will generate waves in its gravitational field. These waves will propagate outwards as ripples in space-time, just as an accelerated electric charge generates electromagnetic waves. There is, however, one difference. Electric charge comes in positive and negative forms, while mass can only be "positive". When a mass is accelerated, there is an "equal and opposite reaction" on the body causing the acceleration, and the motion of the latter cancels out the major (dipole) part of the gravitational radiation. The gravity waves that do propagate are the much weaker, quadrapole, radiation,

---

(Opposite) *The nearest giant cluster of galaxies, the Virgo cluster, which lies some 50 million light years away, may be the source of the highest-energy cosmic rays to bombard the Earth. The collapse of stars in these galaxies may also generate the first gravitational radiation to be picked up by future gravitational wave detectors*

produced most strongly by changes in shape or by the rotation of a non-symmetrical body.

Observations in radio astronomy have hinted that gravitational waves do exist, for a pair of pulsars in close orbit (the Binary Pulsar) is gradually decreasing its separation exactly as predicted if the system's orbital energy is being sapped by the production of gravitational radiation. The collapse of star cores during the explosion of their outer layers as supernovae and matter falling into a black hole should also produce short and more intense bursts of gavitational radiation. American physicist Joe Weber claimed to have detected gravitational waves regularly in the early 1970s, but no one has been able to duplicate his results, even with more sensitive equipment. Now, however, several groups around the world are competing with one another, and with basic limitations set by quantum theory, to achieve a gravitational wave ''telescope'' sufficiently sensitive to pick up the first non-electromagnetic waves from space.

27

Fishing for cosmic rays

ALAN WATSON

28 July 1983

High-speed cosmic ray particles hit the Earth so rarely that astronomers must use special techniques to catch them. The "front end" of the apparatus is the Earth's atmosphere, while the detectors are as diverse as an array of water tanks near Leeds and a cluster of dustbin-sized telescopes in the desert in Utah.

High-energy fragments of matter originating beyond our Solar System bombard the Earth's atmosphere continuously. These fragments, which are mainly atomic nuclei, are called cosmic rays, and the most energetic have energies well above the highest reached by particle accelerators. How Nature arranges to give single atomic nuclei kinetic energies up to 25 joules – energy sufficient to raise a 1-kilogramme weight to a height of 2.5 metres – is one of the outstanding mysteries of high-energy astrophysics. Recent results suggest that the most energetic particles come from outside our Galaxy; if so, high-energy cosmic-ray sources are as powerful as some of the most powerful X-ray and radio sources. But the most energetic particles in Nature are rather rare, and require special techniques to study them.

Cosmic rays play an important role in many aspects of astrophysics, even though we do not know where most of the particles originate and are accelerated. While the most energetic particles may be extragalactic, the bulk of cosmic rays have their origin within our Galaxy. For many years, researchers commonly supposed that cosmic rays were only a rather local feature of extraterrestrial space, confined to a small volume surrounding the Sun. The "local" hypothesis had the attraction that it helped reduce the amount of energy of the observed Universe contained in this mysterious form. But it was overthrown when, in 1938, a French group led by Pierre Auger showed that cosmic rays of at least 10^6 GeV occasionally arrive at the top of the atmosphere. (1 GeV, or gigaelectron-volt, is the energy roughly equivalent to the mass of a proton, and is equal to 1.6×10^{-10} joules.)

Auger made his serendipitous discovery during tests of electronic "coincidence" circuits. He found that when two particle detectors were placed even as far as 300 metres apart the rate of coincident signals in the detectors was much greater than expected by chance. The explanation for this simultaneous arrival of particles at widely separated points was, as Auger showed, that the particles were part of a coherent shower – an extensive air shower of "secondaries" produced by a single high-energy "primary" cosmic ray. The rate of the most energetic showers that Auger observed was about one per square metre per day.

The astrophysical importance of this discovery lay in the support that it gave to a hypothesis of the Swedish physicist Hannes Alfvén, that the Galaxy is permeated by a weak but extensive magnetic field which confines the charged cosmic rays largely to within our own Galaxy. Further evidence that cosmic rays are indeed a feature of all galaxies, came with the advent of radio astronomy. It was then that Vitaly Ginzburg, of the Lebedev Physical Institute in Moscow, and other physicists realised that much of the non-thermal radio emission seen within our Galaxy and from extragalactic sources, could be "synchrotron radiation" – radiation emitted by electrons spiralling in magnetic fields.

Even now, 10^6 GeV is about the highest energy that can be studied directly by detectors flying in balloons or satellites. To study cosmic rays at higher energies, physicists have exploited Auger's discovery of extensive air showers. At 10^{17} eV (1 GeV = 10^9 eV), the radius of the path of a proton spiralling in the galactic magnetic field (the gyration radius) becomes comparable to the size of large-scale features in the Galaxy. Then we can expect variations in the directions from which the cosmic rays arrive, if the particles originate within the Galaxy. At this energy the rate is only one per square metre per century, so extraordinary patience, or large collecting areas, are necessary, and as often happens when planning experiments, economic realities dictate compromise.

A major advance in techniques for studying high-energy cosmic rays came from Bruno Rossi's group at the Massachusetts Institute of Technology (MIT). In the late 1950s Rossi's team developed large-area detectors made of plastic scintillator – a material that emits light when an energetic particle passes through. This fluorescent light is usually in the ultraviolet part of the spectrum, so dye molecules, present as an additive, are used to absorb the ultraviolet light and re-emit it at the longer wavelengths of visible light. Photomultiplier tubes detect this light. They have a light-sensitive cathode that emits electrons which are then "multiplied" by successive electrodes along the tube, converting the light into a pulse of electric charge to be amplified and counted. In a scintillation detector of the type used in cosmic ray work, the scintillating material is housed in a light-tight box with a diffuse reflector so that the light signal at the photomultiplier is nearly independent of the position of the cosmic ray particle within the detector.

Rossi's group showed how these scintillation detectors could be used to measure both the direction of the primary particle and the size of the shower that it produced at the observation level. To see how this is possible it helps to understand some of the characteristics of an extensive air shower.

The shower can be thought of as a disc of particles, shaped rather like a large dinner plate, moving through the atmosphere at the speed of light. The lateral spread is large, a few square kilometres, and arises because particles are emitted at angles from the interactions of the parent particle and its secondaries as they travel through the atmosphere. Scattering of the particles also spreads them far from the axis of the incoming primary particle. The longitudinal spread, along the axis of the developing shower, is small, because the particles nearly all travel at speeds close to that of light. So measurements of the relative arrival times at three (or more) widely separated detectors serves to determine the direction of the shower disc, usually to within 2 or 3 degrees. The size of the shower at ground level can be estimated assuming only that the distribution of particles is circularly symmetric about the shower axis; an incoming particle of 2×10^{19} eV produces about 10^{10} particles at sea level.

The largest array of scintillation detectors built by the group from MIT, at Volcano Ranch, New Mexico, covered an area of 8 square kilometres, with 19 detectors, each of 3.3 square metres, in a pattern of equilateral triangles. This array was operated for over three years by John Linsley, who made the first exploration of the cosmic-ray spectrum beyond 10^{17} eV, detecting about 1000 showers or "events" from primaries of energy greater than 10^{18} eV. In addition to setting limits on the anisotropy – lack of symmetry in arrival directions – of these energetic cosmic rays, Linsley was able to show, by the analysis of a single well-measured event, that the cosmic-ray spectrum extends to at least 10^{20} eV (16 joules), an energy 100 000 times greater than that established in the pioneering French experiments!

No one had recognised the significance of this single event when the first phase of the experiment at Volcano Ranch ended in 1963. But in 1966, Kenneth Greisen from Cornell University and Georgi Zatsepin from Moscow pointed out that the discovery of even one event with an energy of 10^{20} eV, during the observations at Volcano Ranch, was totally unexpected.

In the previous year, Arno Penzias and Robert Wilson, at the Bell Telephone Laboratories in Holmdel, New Jersey, had discovered that the whole Universe is bathed in microwave radiation corresponding to a temperature of 2.7 K – radiation thought to be the relic of the big bang. Greisen and Zatsepin argued that at 10^{20} eV, a proton moving through the 2.7 K radiation field sees the microwaves doppler-shifted to gamma-ray wavelengths. At high enough energies a gamma ray interacts with a proton to produce a nucleon and a pion, and the resultant nucleon (proton or neutron) has less energy than the original proton. The probability for this reaction is so high, and the microwave density so great, that protons with energies above 5×10^{19} eV must interact with the background radiation in less than 10^8 years. This is such a relatively short time on cosmic scales that the observation at Volcano Ranch was quite remarkable. Subsequently it was shown that the same lifetime limit holds if the 10^{20}-eV cosmic rays are heavy nuclei, such as iron, in which case photo-disintegration reduces the energy of the nuclei.

Of course it was possible that the single event seen at Volcano Ranch had been an unusual fluctuation. So, the search for a high-energy "cut-off" in the cosmic-ray spectrum motivated researchers to develop large-area arrays in England, Australia and the Soviet Union. The groups were also interested in measuring the pattern in the arrival direction of the particles and the masses of the most energetic particles. The British array, built at Haverah Park, 28 kilometres north of Leeds, by John Wilson's group from Leeds University, was brought into full operation during 1968. Groups from Imperial College, London, and the Universities of Durham and Nottingham developed supporting experiments to study detailed aspects of showers, and to help solve the question of the masses of the particles.

The concept of the array at Haverah Park differs somewhat from the one at Volcano Ranch, and was partly dictated by the available land. Instead of a regular lattice of small detectors, the team at Leeds constructed large-area detectors at a small number of sites. To reduce cost, the detectors were made of water, with 600 000 litres of local water filling 550 square metres of detector, and the charged particles were detected through the Cerenkov effect.

In 1934, the Russian physicist, Pavel Cerenkov, had discovered that when a charged particle passes through a dielectric medium at a speed greater than the velocity of light in that medium, it generates a continuous spectrum of emitted light along its track. (An acoustic analogy is the shock wave produced when an aeroplane flies faster than the speed of sound.) Once photomultipliers became readily available the Cerenkov effect was widely used for the detection of charged particles.

The detectors used in the cosmic-ray air-shower array at Haverah Park exploit the Cerenkov effect in water. Each tank (2.25 square metres in area and 1.2 metres deep) is filled with water and lined with white plastic which diffuses the Cerenkov light, some of which is collected by a photomultiplier with a 12.5-centimetre diameter photocathode. Some 250 photons/centimetres are emitted at visible wavelengths, approximately 0.5 per cent of which are detected.

The layout at Haverah Park, covering about 12 square kilometres is apparent from figure 1, which shows the density pattern of a large, well-measured "event". The wide range of particle densities, from more than 8000 down to 0.4 per square metre is typical of a giant air shower. It is clear that we can determine the densest region of the shower with precision, assuming only circular symmetry. We have estimated the energy of the cosmic ray that produced this density pattern as 1.0×10^{20} eV. The estimate is based on computer calculations made by Michael Hillas, from Leeds University, in which he incorporates contemporary data from experiments at particle accelerators. The calculations predict many detailed features of air showers rather well, so we are confident that the estimates of energies are not grossly in error, even if we make extreme assumptions about the masses of the primary particles, and despite the wide range of assumptions we can adopt to describe the physics of the nuclear interactions.

An experimental check on the energy calibrations comes from measurements, particularly those of the Soviet group, of Cerenkov light produced in the air above the detector.

As a cosmic-ray shower develops, the more energetic charged particles emit Cerenkov light in the air. Because air is less dense than water fewer photons are emitted by each particle per unit length than in the water-filled detectors (0.3 photons per square centimetre at sea level) but the large number of particles leads to a signal that is

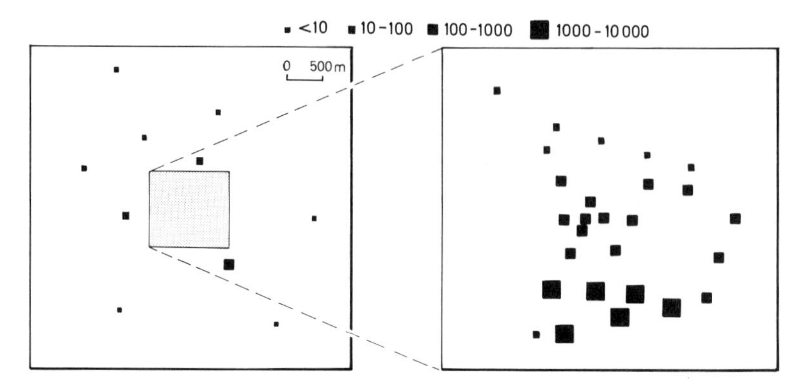

Figure 1 *The density pattern of particles per square metre in an air shower produced by a 10^{20} eV primary cosmic ray, recorded by the array at Haverah Park. The detector areas vary from 34 square metres down to 1 square metre at most of the positions in the central region, which is shown enlarged*

readily detected at ground level on clear moonless nights. The Cerenkov light is emitted in a cone of angle θ, where, for very fast particles, $\cos \theta = 1/n$, where n is the refractive index. The Cerenkov angle is about 41 degrees in water but only about 1.3 degrees in air at sea level. The light from the air shower is thus very directional, but we can detect signals out to at least 1 kilometre because electrons in the shower are strongly scattered.

As the particles in the shower pass through the atmosphere they dissipate about 80 per cent of their total energy as they ionise the air. Cerenkov light emitted simultaneously allows us to estimate the energy lost through ionisation and, when coupled with measurements of the energy retained by particles in the shower when they reach the level of the detector, we can estimate the primary energy. Cross-checks of independent estimates of energy made at Haverah Park, Volcano Ranch and Yakutsk (in the Soviet Union) have been made using scintillation detectors common to all three arrays: these show that even at 5×10^{19} eV the agreement is excellent, within 15 per cent.

The array at Haverah Park has recorded about 200 showers produced by primaries of 10^{19} eV and above. The rate of events over 10^{20} eV is only a few per square kilometre per century; and the spectrum extends to at least 1.5×10^{20} eV. The spectrum, although not well determined from so few events seems to have a flatter slope above 10^{19} eV than between 10^{17} and 10^{19} eV. However, what is certain, is that cosmic rays arrive at Earth with energies well beyond the cut-off at around 5×10^{19} eV that Greisen and Zatsepin predicted. It follows that the most energetic particles are rather young, on a cosmological time scale, probably less than 10^8 years old. Moreover, there is no evidence that the cosmic rays arrive from directions clustered about the region of the galactic plane as we might expect if the particles of the highest energy are protons *and* their sources lie within our Galaxy.

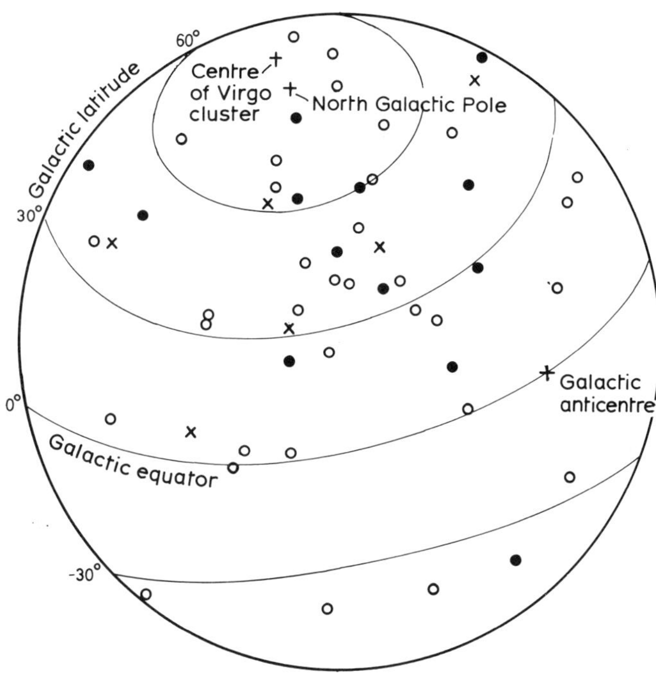

Figure 2 *The directions from which the highest energy cosmic rays arrive are here plotted on a map of the northern sky. These 54 cosmic rays, with energies greater than 4×10^{19} eV, were recorded at Haverah Park, UK (O), Volcano Ranch, US (×) or Yakutsk, USSR (●). The "Galactic equator" is the line of the Milky Way, as seen in the sky; 90 degrees away from it is the North Galactic Pole. The nearest large cluster of galaxies, the Virgo cluster happens to lie in almost the same direction. There is clearly no indication that these cosmic rays originate in the Milky Way galaxy. Their arrival directions indicate they come from beyond, perhaps spreading out from the Virgo cluster*

To tie down the origin of the most energetic cosmic rays we need some idea of what masses they have. This is a tough problem, but a new application of the technique of detecting Cerenkov light from the atmosphere promises to supply an answer. The depth in the atmosphere of the maximum number of particles in the shower reflects the energy and mass of the primary particle: more energetic particles are more penetrating, so the shower develops at a lower level. We can infer the level of the maximum from the temporal and lateral spread of the Cerenkov-light signal seen at ground level. Ted Turver's group, at Durham University, has operated an array of Cerenkov-light receivers in the Dugway Desert, west of Salt Lake City in Utah (the sky at Haverah Park is rarely clear enough for such observations!). Data from this array, and others, suggest that the masses of cosmic rays with energies in the range 10^{17}–10^{18} eV are significantly lighter than at lower energies. Perhaps 65 per cent of these cosmic rays are protons. While this conclusion is not yet universally accepted, it agrees with independent estimates made at Haverah Park, where the group from Leeds has exploited the large areas of the detectors in related studies of the temporal spread of shown particles. Furthermore, the data show that iron does not dominate the beam above 10^{19} eV. This is particularly important because iron nuclei originating in our Galaxy might just be consistent with the spectrum and directions of arrival that we observe.

It seems justifiable to speculate that the most energetic particles are in the main protons, and thus of extragalactic origin. Assuming that these have their origin within the local super-cluster of galaxies (radius around 60 million light years), a limit set by the estimated lifetime, we reach the important conclusion that the observed power at the source of these cosmic rays is 3×10^{35} watts. This is comparable to the output of the sources Centaurus A or M87 in the radio band or NGC 4151 in X-rays. Ultra-high energy particles are a major drain on the cosmic energy budget!

If the most energetic cosmic rays are truly the only sample of extragalactic matter to reach us, it is important to discover more about them, as they must provide clues to the nature of the acceleration region. Over 20 years after Linsley's first detection of a 10^{20}-eV cosmic ray, the world total of such well-measured events is still only about 10! The very low flux demands that we think of new techniques, because 10–20 square kilometres is about the maximum that can be exposed with the methods I have described so far.

Nearly 20 years ago Greisen proposed the detection of scintillation light from atmospheric nitrogen as a method for observing air showers from 10^{19}-eV cosmic rays. Laboratory tests showed that nitrogen scintillates only weakly (only one-hundredth as brightly as the Cerenkov radiation from air), producing about 5 photons per metre of an electron track, at wavelengths between 310 and 410 nanometres. But it still seemed possible to detect the light emitted equally in all directions at large distances from the shower, and so "see" the shower side-on. Early estimates indicated that, at a site with good viewing conditions, it might be possible to detect 10^{20}-eV showers from a distance of up to 50 kilometres. This would correspond to an annual exposure of more than 700 showers per square kilometre. Moreover, if the time of arrival of the scintillation

Two of the 67 detectors of the American "Fly's Eye". In background at right, one points vertically downwards. The detector at left reveals its 1.5 metre diameter mirror, reflecting an inverted view of the desert. At the mirror's focus is an array of photomultiplier tubes to detect the faint light produced by a cosmic ray travelling through the atmosphere

light was measured, it looked feasible to trace directly the growth and decay of the shower as it developed through the atmosphere.

Early attempts in Australia, Japan and the US to "see" showers through atmospheric scintillation light were partially successful but fell short of fulfilling the promise of the technique, mainly because of limitations set by the electronics then available. In the mid-1970s a group from the University of Utah began a new attempt to realise Greisen's detector, which he dubbed the "Fly's Eye". This device consists of 880 light-sensitive photomultipliers, each 76 millimetres diameter, mounted at the foci of 67 spherical mirrors, diameter 1.5 metres. The mirrors are arranged to cover a complete hemisphere of the sky, and the instrument is mounted on a granite outcrop in the Dugway Desert, western Utah. At an early stage the group ran three "mirror units" in coincidence with the array at Volcano Ranch and established that the showers from 10^{18}-eV primaries could be detected by their atmospheric scintillation light.

The Fly's Eye, in its full form, has now begun to gather data on air showers; it has demonstrated its ability to measure accurate shower profiles and has recorded a few events above 10^{19} eV. With the electronics presently installed, the Eye has not been able to see further than about 10 kilometres; background light, some of it Cerenkov light from small showers scattered by aerosols, has proved to be a problem. To improve the perception of depth – and hence the accuracy of the measurements – the team at Utah has commissioned a second, smaller, Fly's Eye. The combined system will investigate the energy spectrum of cosmic rays to beyond 10^{19} eV, and provide new information on the mass composition and nuclear physics properties, in particular the interaction rates between protons and air above proton energies of 10^{18} eV.

Other schemes to boost the rate of recording events above 10^{19} eV and to explore the decade beyond 10^{20} eV are being examined. Linsley has proposed mounting apparatus like the Fly's Eye on a satellite. He visualises a mirror of 36 metres in diameter with 5000 photomultiplier tubes at its focus. The phototubes would monitor 10^{11} tonnes of atmosphere and would be largely free from problems of variable visibility which trouble similar ground-based experiments. Koichi Suga and his group at the Tokyo Institute of Technology plan to exploit the fact that the dense core of protons and related particles at the centre of an ultra-high-energy cascade generates an acoustic (sound) signal in any homogeneous medium through which it passes. The attenuation of this signal can be very small near frequencies of 10 kHz, so an array of well-separated hydrophones in a lake might detect the highest-energy cosmic rays by listening for them. Unfortunately exploratory experiments at a possible site, Lake Titicaca in Bolivia, suggest that the background noise level there may be too high to allow the technique to be developed economically.

Another approach is to extend still further the scale of conventional particle-detector arrays. A system linked by cables becomes impractical for an area greater than about 10 square kilometres, so the group at the University of Sydney developed an alternative during the 1970s. The team set up a system of 50 autonomous detector-pairs in the Narrabri Forest in New South Wales, on a grid covering about 50 square kilometres. Tape recorders store the times and amplitudes of signals in twofold coincidences at each pair, the absolute time being derived from a signal transmitted across the array. A computer subsequently identifies coincidences between widely spaced stations. When the final analysis of this experiment is completed it will provide a survey of the cosmic-ray spectrum and arrival directions, from 5×10^{17} to 10^{20} eV, as seen from the Southern Hemisphere.

This experiment was conceived too early to take advantage of cheap microprocessors and memory, and the researchers had to bury the detectors to keep the rate of coincidences between pairs of detectors acceptably low; the ground absorbs electrons in the cosmic-ray showers, leaving the muon component alone to be detected. But the basic concept points the way to the construction of the next generation of arrays. To explore the cosmic-ray spectrum beyond 10^{21} eV it will be necessary to build a 1000 square-kilometre array, preferably at a site on the Equator. Such a project would be expensive, but much less costly than a satellite project, and might be set up by an international collaboration in the 1990s. In the meantime, techniques must be developed to record and transmit data from a large number of autonomous stations. Such work is in progress in Japan, where the Institute for Cosmic Ray Studies is extending its existing array from 1 to 25 square kilometres and in England, where at Leeds University we have plans to extend our 12 square-kilometre layout to about 100 square kilometres. The cost of the expansion at Leeds will be minimised by re-developing the existing detectors on a 1-kilometre triangular grid. The array will embody some of the principles of the Australian experiment but, in addition, modern techniques make it feasible to control the stations remotely and to transmit data from them to a central computer.

In the next decade experiments such as the Fly's Eye will consolidate our knowledge up to 10^{20} eV, and Fly's Eye and a new generation of gigantic arrays will surely explore the region beyond. The end of interest in the most energetic particles in Nature, like the end of the cosmic-ray spectrum, is not yet in sight.

28

Neutrinos in the swim

"MONITOR"

26 June 1980

A bold plan to detect the first neutrinos from beyond the solar system would involve a cubic kilometre of the Pacific Ocean as a detector.

An international group of scientists plans to study neutrinos, elusive neutral particles that can pass straight through the Earth with extremely high energy, by detecting the small proportion of neutrinos that are stopped in a large volume of sea water which contains a three-dimensional matrix of particle detectors. The apparatus would be sited off Hawaii, and is expected to cost several million dollars.

Called DUMAND (deep under-water muon and neutrino detector), the apparatus would be used by scientists in a variety of disciplines. Particle physicists would use it to study the interactions of energetic neutrinos with matter, for DUMAND will pick up neutrinos that are far more energetic than those that present particle accelerators produce. Astrophysicists could use the detector to find out where the neutrinos come from in space, and how many there are. They believe the information will add to their knowledge of the energetic processes that fuel active galaxies and pulsars.

DUMAND could even detect bioluminescence from sea creatures, for the apparatus is designed to register light – specifically the Cerenkov radiation that comes from energetic electrically charged particles which pass through the sea water.

According to Arthur Roberts of the Hawaii DUMAND Center at the University of Hawaii, the detector will consist of some 1000 light-sensitive "sea urchins" placed at intervals of 40–50 metres throughout the 1 cubic kilometre volume of water. Each urchin will have 3-metre long "spines" which are narrow tubes that fluoresce when they absorb Cerenkov radiation. The fluorescence will be detected by a light-sensitive photomultiplier tube at the heart of the "urchin's" hemispherical body.

29

The hunt for gravitational waves

JAMES HOUGH and RONALD DREVER

17 August 1978

Gravitational radiation should be reaching us all the time from space, but the construction of a "telescope" to detect these ripples in space-time is one of the most difficult problems in physics today.

Albert Einstein predicted in 1916 that gravitational radiation, a wavelike disturbance of space travelling with the velocity of light, and carrying energy, should be produced when matter accelerates in a suitable way. Significant amounts of this radiation are expected to be generated during the collapse of stars, the production and interactions of black holes, and other violent astronomical processes. The detection and analysis of the resultant gravitational waves could prove a most valuable tool in astronomy. It could give information obtainable in no other way on events occurring in the cores of stellar systems surrounded by great depths of screening material which cannot be penetrated by light, X-rays or other kinds of radiation. It could also be expected to provide important clues on the existence, nature and behaviour of black holes. Measurements of the polarisation and velocity of gravity waves should also provide valuable data to help differentiate between rival theories of gravitation. In view of these potential rewards, physicists are investing a considerable effort around the world in developing gravitational wave detectors of ever-increasing sensitivity. In this article we shall discuss some of these devices.

How can gravitational waves be detected? Essentially by their property of moving neighbouring pieces of free matter relative to each other. These forces act at right angles to the direction in which the wave is travelling, and the movement produced (εL) is proportional to the separation of the masses (L). This makes it convenient to describe the amplitude of a gravitational wave at the

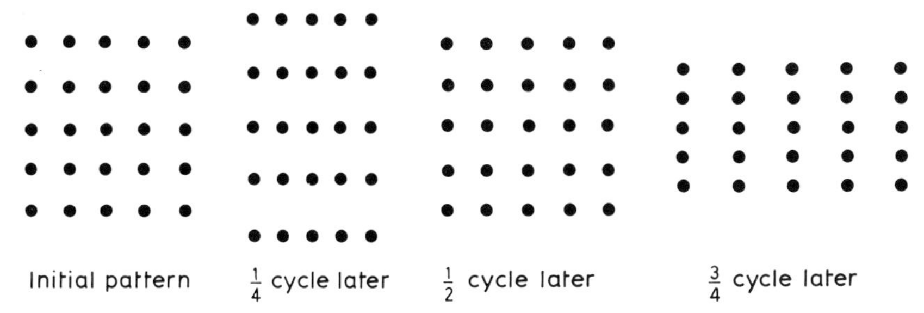

Figure 1 *Gravitational waves make an array of test particles oscillate alternately in the two directions at right angles to its motion. Here, the wave is travelling into the page, and the particle positions are indicated as one cycle of the radiation passes through*

detector in terms of the value of $\delta L/L$ induced there in a system of free particles (Figure 1).

In most of the gravity wave experiments performed so far researchers have looked for signals with a predominant frequency in the region of 1 kilohertz (1000 cycles per second). A typical detector consists of a single aluminium bar for which longitudinal vibrations have a resonant frequency of this order. For short pulses with most of their energy in waves of frequency close to the resonant frequency of the bar, the two halves of the bar behave rather like two free test masses. The motions induced can then be detected by measuring the strain set up in the bar, which is approximately equal to $\delta L/L$.

Theoretical estimates of magnitudes of signals expected at the Earth from some possible sources of gravitational radiation are given in Table 1. From this two factors become clear. First, the effects to be measured are extremely small. Even for stellar collapses in our own Galaxy, for example, the motion expected at the ends of a detector 1 metre long is less than the diameter of the nucleus of a hydrogen atom. And in order to "see" out as far as the Virgo cluster, where an interesting rate of stellar collapses is expected, unprecedented sensitivity in measurement is required. Secondly, it seems important to extend the frequency range of the detectors from the kilohertz band of the early experiments to cover lower frequencies where a variety of other possible sources may give signals.

In order to understand how improvements in the sensitivity of detectors can be made, we should consider some of the limiting factors. One fundamental factor is thermal motion, the random motion of atoms in the bar due to their "heat energy". In a resonant

Table 1 Estimates of gravitational radiation from some possible sources

	Predominant frequency (hertz)	$\delta L/L$ *at Earth*	*Event rate*
Stellar collapse*			
Our Galaxy	10^{2}–10^{4}	10^{-19}–10^{-17}	~ 1 or more per 30 years
Virgo cluster	10^{2}–10^{4}	10^{-22}–10^{-20}	~ 10 or more per year
Black hole events			
in globular clusters	1–10^{2}	10^{-19}–10^{-21}	up to 10 per year
in galactic nuclei and quasars	10^{-4}–10^{-1}	10^{-16}–10^{-19}	up to 50 per year
Pulsars			
Crab Nebula pulsar	60	~ 10^{-27}	continuous
Vela pulsar	22.5	~ 5×10^{-26}	continuous
Binary stars			
Iota Bootes	8.7×10^{-5}	5×10^{-21}	continuous
AM CV_n	1.9×10^{-3}	10^{-21}	continuous
Total from all binaries in our Galaxy	~ 10^{-5}–10^{-2}	10^{-19} to 10^{-20}	continuous

* Significant radiation at a few hundred hertz may come from neutron star binaries produced in stellar collapse

bar detector this shows up as a slowly varying oscillation at the resonant frequency. Now the effect of a pulse of gravitational radiation is to cause a change in either the amplitude or phase of this motion. Since this change has to be observed against the thermal background it is clear that the smaller the background, and the more slowly it fluctuates, the better. The size of the background depends on the mass of the detector (the larger the mass the smaller the relative background) and on temperature. The rate of fluctuation increases with the degree of damping of oscillations in the detector – so the less damping the better. We shall return to this point later.

Another, even more fundamental, factor which might limit the sensitivity of a gravitational wave detector arises from the Heisenberg Uncertainty Principle. This says it is impossible to know simultaneously both the position and the momentum of a particle with perfect accuracy. Indeed, if one attempts to determine very precisely the position of an object, such as one of the test masses of a gravity wave detector, one is forced to disturb its momentum in a way which is unpredictable and could not be distinguished from the effect of a gravitational wave. Hence the sensitivity of the detector is restricted. A similar type of argument can be applied to a resonant-bar gravity wave detector. In this case you can show that the weakest gravity wave detectable by simple displacement measurements is that which would feed in energy equal to one quantum at the resonant frequency of the bar, if the bar initially had zero energy. (Similar arguments have long been known to apply to electrical amplifiers, and set an ultimate limit to performance of such practical devices as maser amplifiers used in microwave communication.)

Fortunately, recent suggestions indicate some ways of getting round this problem. Direct measurements of the energy of the bar may escape from the quantum limit if they are made relatively slowly. However, details of the gravity wave signal may be lost. Another possible solution has come from very recent work at California Institute of Technology by K. S. Thorne and colleagues, and one of the present authors. If the coupling between the bar and the measuring apparatus is modulated periodically at the same frequency as that at which the bar resonates, then the disturbing effects of the measurement occur out of phase, so making them unimportant. One may do better still with a pair of modulated sensors, measuring the velocity of the end of the bar and the amplitude, and may beat the quantum limit with less loss of information about the wanted signal.

At present, however, practical realisation of such schemes is not going to be easy – in current experiments other sources of noise are more serious than the quantum limitations – but it looks as though in the future they may become of vital importance. These new ideas will probably have applications in other fields, such as measurement of very low-frequency electromagnetic signals, and spin-off from gravity wave research may well have real practical value elsewhere in the near future.

The first approach to improving detector sensitivity is to increase the mass of the aluminium bar and cool it to very low temperature, thus reducing the thermal noise. Equipment of this type is being developed at Stanford University by W. M. Fairbank, at Louisiana State University by W. O. Hamilton, and at the University of Rome by a group led by E. Amaldi and G. Pizzella (Figure 2).

The aluminium bars in this series of experiments weigh up to 6 tonnes and eventually they may be cooled to temperatures of the order of 0.05 K. However, there is great difficulty in measuring the very small motion of the bars without introducing a considerable amount of additional noise. Various sensors are being developed to deal with the problem. One uses a resonant superconducting diaphragm coupled to the bars. Shaking the diaphragm modulates the inductance of flat superconducting coils placed close to it. The inductance change is then detected by a superconducting amplifier – a Josephson junction magnetometer.

Unfortunately the performance of these and similar sensors is not as good as one would like. Their lack of sensitivity may limit this type of detector in the immediate future to below the sensitivity which would be required, for example, for detection of stellar collapses at the distance of the Virgo cluster. However, active development of sensors is continuing and improvements can be expected, perhaps based on some of the new ideas mentioned above.

Another interesting approach has been proposed by V. B. Braginsky and his group at Moscow University, and is being investigated by them and by groups led by D. H. Douglass at the University of Rochester and J. Weber at the University of Maryland. This is the use of a single crystal of sapphire or silicon as the resonant element in the detector. Theory suggests that perfect crystals of sapphire may have exceedingly low damping at a temperature of a few degrees above absolute zero, and already damping times of the order of days have been measured by the Moscow group. This means that if such a crystal were struck a blow to set it ringing it would carry on oscillating for this time before its amplitude fell significantly.

With such long damping times, effects of thermal noise are significantly reduced, and this may compensate for the relatively small mass of available single crystals (although 100-kilogramme sapphire crystals are expected to be manufactured within the next few years). Of course, there is still the severe problem of sensing the motion in these detectors. Braginsky and his colleagues have plans to let the movement of the crystal bar modulate a superconducting microwave cavity and eventually to detect the modulation by a technique similar to those mentioned above for sidestepping quantum limitations (Figure 2). This new technology of ultra-low-loss crystals shows great promise for sensitive physical measurement of various kinds. Within the next decade detectors of this kind

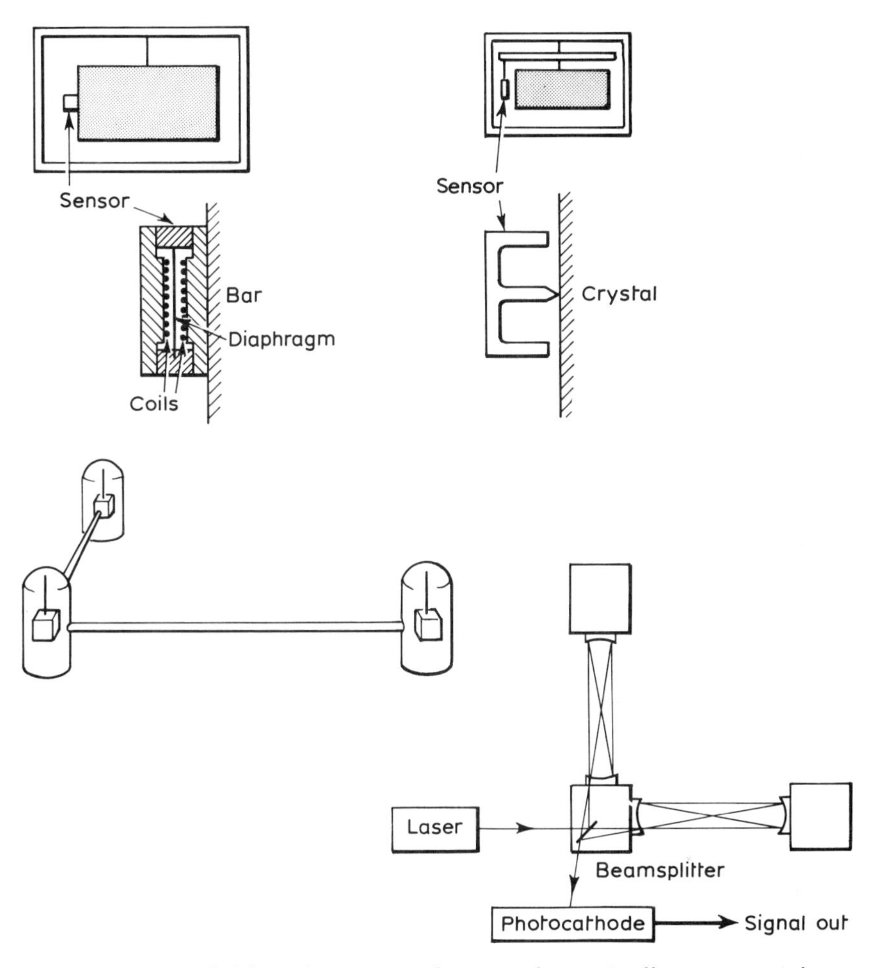

Figure 2 *Solid bar detectors, shown schematically at top, pick up one component of the oscillations due to a gravitational wave. At top left is a detector incorporating a heavy aluminium bar; its motion can be measured from the change in inductance in the two superconducting coils, when the bar's oscillations shake the superconducting diaphragm between them. The detector design at top right incorporates a large sapphire crystal, whose motion is measured by the altering size of the cavity in the sensor, which affects the frequency of microwaves resonating within it. The long baseline interferometer (below) measures the difference between the motions induced in two perpendicular directions. Laser light travelling along each arm and reflected back is recombined to form interference fringes; if the relative paths are changed by the effect of gravitational waves, the fringes will alter at the same rate*

may come close to allowing the detection of gravitational radiation from stellar collapses in the Virgo cluster.

A different method of improving the sensitivity of a gravitational radiation detector is to increase the displacement caused by the wave rather than attempt to reduce background effects to extremely low levels. This may be done by making the separation L between a pair of test masses large. P. S. Aplin, at the University of Bristol, is building a detector with a distance between the test masses of nearly 50 metres. He intends to couple the motion of the masses to a piezoelectric sensor by means of a horn-shaped acoustical transmission line.

For masses at large separation, optical interferometry with a laser comes to mind as a promising method for detecting changes in separation. It has the important advantage of giving negligible damping, so that thermal noise can be very small even at room temperature. In order to avoid effects of wavelength fluctuations in the light source used one may look for relative changes in two perpendicular baselines. One may use three masses, each possibly suspended like a pendulum, with mirrors attached to form a Michelson interferometer. Gravitational radiation will cause changes of opposite sign in the two path lengths and small, but possibly detectable, changes in the optical interference pattern. This approach was first tried by R. Forward of the Hughes Research Laboratories. With separations of 3 metres he could detect movements of about 10^{-15} metre, which is less than 10^{-8} of the wavelength of light (Figure 2).

Further, as suggested by R. Weiss of MIT, the change in path length caused by a gravitational wave can be increased by reflecting the beam of light in each arm of the interferometer back and forward many times between a system of concave mirrors. With high-reflectivity dielectric mirrors several hundred reflections can be achieved in each arm of the interferometer. Detectors of this type are being constructed at MIT, at Glasgow University and by H. Billing and colleagues at the Max Planck Institute for Astrophysics in Munich. One of the important properties of this type of detector is that with masses suspended like pendulums the frequency range extends from a few hertz to a few kilohertz which is much wider than that of the resonant type of detector. This allows coverage of some of the lower frequencies of interest.

In order to estimate the sensitivity of these instruments it is necessary to consider a number of factors. These include statistical fluctuations in the number of photons detected in the interference pattern, statistical fluctuations in the radiation pressure on the mirrors on the test masses, thermal noise from the suspension of the test masses and in the masses themselves, Brownian motion caused by residual gas in the vacuum system enclosing the whole apparatus, and seismic noise. An analysis of these factors for a detector, using 150-kilogramme masses separated by 1000 metres and illuminated by a laser of 2 watts power, yields the encouraging result that pulses from black holes in globular clusters, whose energy spectrum peaks between 1 and 100 hertz, may be detectable. Further, given an increase in available laser power, such detectors may have the sensitivity to allow stellar collapses in the Virgo cluster to be detected. Thus it seems possible that these detector systems will open a broad field of experimental investigation. There are, however, many technical problems to be overcome, and active work on these is going on.

At present at Glasgow University we are developing techniques for detectors using optical interferometry in a system with two 300-kilogramme aluminium masses whose centres are 1 metre apart. This apparatus, based largely on components from previous gravity wave experiments by W. D. Allen at Reading University, is enabling the testing of some new ideas for the optical arrangement and suspension of the masses and will soon be followed by a larger system with 10-metre baselines. We anticipate further increases in baseline to the order of a kilometre in the future.

To search for gravity waves at even lower frequencies, where supermassive black holes may radiate, it would be appropriate to use still longer baselines. As seismic disturbances and local gravity fields of moving objects are more serious at these frequencies, there could be advantages in using a free-mass system involving one or more interplanetary spacecraft as well as the Earth. The doppler tracking radio signal used for navigational purposes provides a possible method of sensing the interaction of gravitational radiation. The main limitations to sensitivity here come from the effects of the interplanetary medium on the radio signal, the stability of the master clocks used, and buffeting of the spacecraft by the solar wind. Improvements may result from using multiple radio frequencies for the tracking, better clocks, and "drag-free" satellites in which an internal screened free mass controls the orbit of the spacecraft. In the longer term, laser tracking between separate spacecraft may give the sensitivity required for detecting even the weak gravity waves predicted to come from binary stars. Such systems would almost certainly detect gravity waves from many other sources also.

Detection of gravitational radiation probably presents one of the hardest challenges to the experimental physicist at the present time – but we hope we have made it clear that several different ways of meeting the challenge look promising. As well as the exciting new window on the Universe that should open with detection of gravitational radiation, the study of ultrasensitive measurement involved has already led to interesting spin-off. The problems are difficult – but there is a feeling that one can now see how they may be resolved, and this is reflected in enthusiasm and excitement among workers in the field.

PART SEVEN

Optical Astronomy: Space to Expand

The excitement of the "new astronomies" can easily overshadow the "old astronomy" – the study of the Universe by the light that reaches us from the skies. The impression is strengthened by the fact that the world's largest optical telescope (barring a Soviet colossus that has never fulfilled its potential) is still the 5-metre (200-inch) Hale Telescope on Palomar Mountain in California. This telescope was opened in June 1948, before astronomers knew of any other radiation from space except radio waves, and radio astronomers had still to identify any radio sources apart from the Sun and the Milky Way.

But this view is deceptive, for optical astronomy is right at the forefront of modern research. The results from radio telescopes, X-ray telescopes and their kin are difficult to interpret unless there is an optical identification of the radio or X-ray source. At the simplest level, an optical picture of that region of the sky – a photograph or television image – can show whether the object is a star, a nebula or a galaxy. Optical spectra can then reveal much more, building on the firm basis of a century's analysis of the light from such objects.

Optical astronomers are usually able to provide the most crucial piece of information about any newly discovered object: its distance. Many X-ray sources, for example, are streams of hot gas within a double star system where gas from an ordinary star is falling towards a compact neutron star or black hole. The X-ray data provide no clue to the system's distance, not even whether it lies in our Galaxy or in a galaxy beyond. But optical astronomers can obtain the spectrum of light from the ordinary star, and hence deduce its type. Knowing the intrinsic brightness of the star, they can calculate its distance. Provided with the distance, X-ray astronomers can then calculate how powerful the system is at the wavelengths they study. In such double star systems, incidentally, optical astronomers can provide another crucial determination. By observing the doppler shift in the

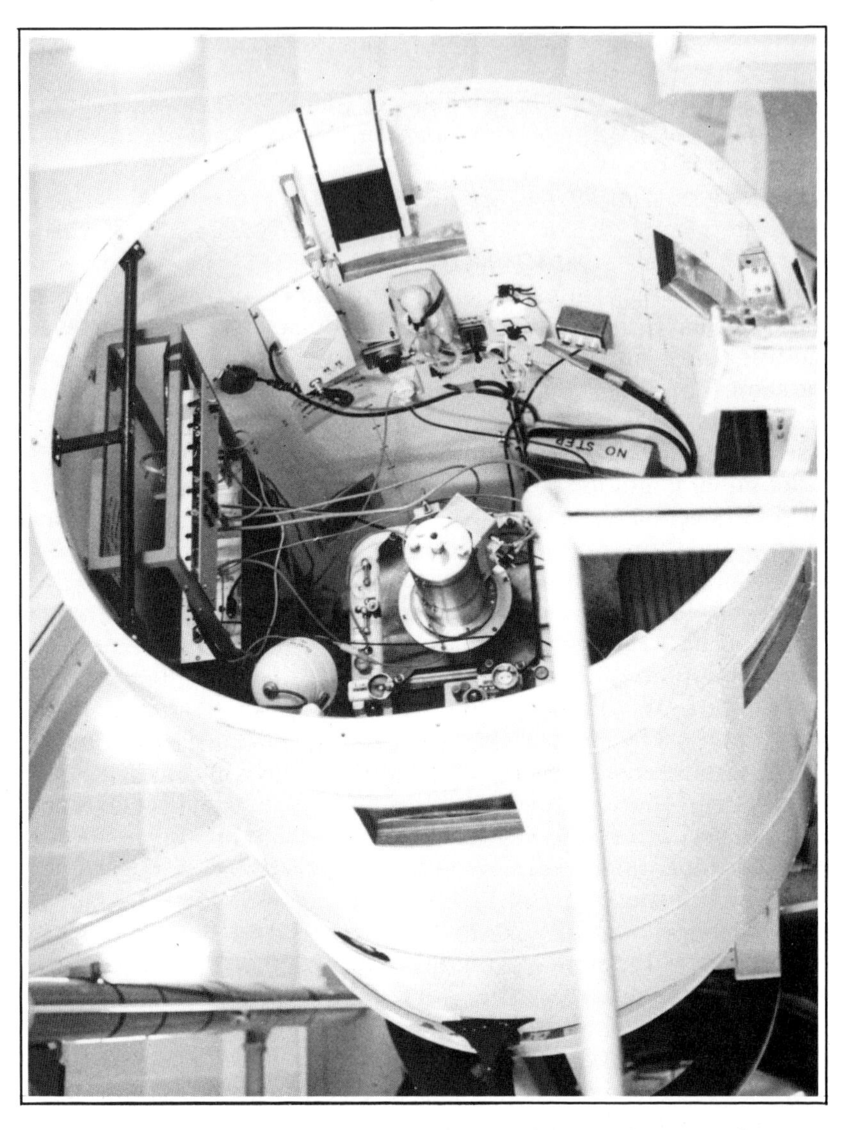

Years ago, a warmly clad astronomer would sit in the prime focus cage of a giant optical telescope, controlling the starlight falling onto a photographic plate. Now his place is taken by a far more sensitive charge-coupled device (CCD) camera, seen in its liquid nitrogen-cooled jacket (centre) at the prime focus of the 3.9-metre Anglo-Australian Telescope (see p. 204). A CCD camera can automatically achieve in minutes what it took an astronomer with plate camera many hours to record

spectral lines of the optically detected star, they can calculate the mass of its compact unseen companion, and work out whether it is a neutron star, or a black hole.

Looking out farther into space, many radio and X-ray sources are associated with distant galaxies, quasars or clusters of galaxies. Optical astronomers have established a "ladder" of distance determinations, using a series of standard indicators such as the brightest star in a galaxy, the size of the largest nebulae in a galaxy, and even the brightness of particular types of galaxy that always seem to have the same luminosity. This ladder extends out to over 1000 million light years, and a galaxy's distance can be ascertained directly if it is associated with a galaxy whose distance is known from this ladder. A more general method relies on the fact that the Universe is expanding, so that clusters of galaxies are thus all receding from one another. The expansion stretches the radiation from a distant galaxy, and so all its spectral lines are moved towards the red end of the spectrum – they are *redshifted*. The farthest and most powerful objects, including radio galaxies, quasars and the gas in rich cluster of galaxies, produce no spectral lines at radio wavelengths and their X-ray lines cannot be detected by present-day X-ray telescopes. But an optical spectrum will show several lines from a galaxy or quasar, and so instantly reveal the redshift – and hence the object's distance.

Optical astronomy also has the advantage of being cheap, even when observatories are placed on distant mountain tops to achieve the best view of the sky. The new William Herschel Telescope at the international observatory on La Palma, for example, will be the world's fourth largest optical telescope, and second to none in performance. Yet it will cost only £10 million. This is less than one-fifth the cost of America's Very Large Array radio telescope or of even a very moderate astronomy satellite. For the price of the pioneering infrared satellite IRAS, one could build 10 William Herschel Telescopes – or a ground-based optical telescope over 25 metres in size, five times larger than the Palomar telescope. And a ground-based telescope has a virtually unlimited life, as witnessed by the Palomar 5-metre which is still one of the world's leading instruments 35 years after it was built. Most satellites, despite their expense, can operate for only two or three years before they run out of control gases, their batteries or solar panels deteriorate, or – for infrared satellites – their liquid helium coolant boils away.

The lack of a competitor to the 5-metre telescope (apart from the Russian 6-metre) does not mean that optical astronomy has stood still during the revolution of the past 25 years. In fact, the period has seen the opening of seven new telescopes larger than 3 metres in aperture, and many smaller instruments. They are so much in demand that less than half the applications for observing time on the big telescopes can be accommodated.

Telescopes in the size range 2.5–5 metres are in fact very efficient collectors of light, but traditional methods of detecting light have wasted most of the radiation when it reaches the telescope's focus, and even the light that is recorded has been analysed in inefficient ways. Up to the middle of this century, astronomers recorded the telescope's view on a photographic plate, and then analysed the plate – whether a direct view or a spectrum – by hand, often by the tedious process of peering through a travelling microscope, reading off the microscope's motion, and then calculating in a laborious manner the corresponding positions in the sky or wavelength in a spectrum. This method wastes most of the information in a photographic plate, particularly if it is a wide-angle picture taken with a Schmidt telescope and containing the images of millions of stars and galaxies.

Now, automatic plate-scanning machines can look through a whole plate without human intervention, measuring positions and brightnesses of stars and galaxies and storing them on magnetic tape for the astronomer to retrieve in whatever form he wants – for example, as a graph of the number of galaxies of different magnitudes. A few hours with a plate-scanner and computer can achieve what would previously have taken literally years of work.

But the photographic plate is a particularly poor way of recording information (except when a large region of sky must be recorded). Even the best photographic emulsions respond to only a small percentage of the light falling on them. Electronic detectors are far more efficient. The first were photomultiplier tubes, which came into use in the 1940s. They could record the brightness of just one object at a time, but the development of sensitive image tube systems has meant that astronomers can now form intensified images of the view or spectrum at a telescope's focus. The most sophisticated image tube system, the Image Photon Counting System, uses a television camera to look at the image tube's output screen. The camera is attached to a computer "brain" that can decide whether each spot on the screen is produced by a photon of light,, or is noise in the tube. It ignores the latter, to produce a "cleaned-up" image.

More sensitive, though, and more versatile, are the solid-state devices called CCDs (charge-coupled devices), large silicon chips sensitive to light. Their surfaces are divided into many individual light-sensitive regions, each of which builds up an electric charge

The CCD (charge-coupled device) is a highly-sensitive light-detecting silicon chip, its surface divided into hundreds of thousands of individual "pixels" or picture elements. In a modern telescope, all the light collected by a huge mirror is focused onto this small detector – note the finger-tips for scale.

proportional to the intensity of light falling on it. After an exposure of a few minutes (equivalent to several hours with a photographic plate), the charges are "read out". This electronic image can be stored in a computer, displayed on a screen, and manipulated to bring out details and information that the astronomer wants. The great advantage of CCDs is that they respond to over 70 per cent of the light falling on them, closely approaching the theoretical ideal of 100 per cent efficiency.

Optical astronomy is thus experiencing a boom, as it complements the other astronomies, and explores new territory itself. But although ground-based telescopes are cheap and easy to use, they have two disadvantages. The sky is never completely dark, because even away from artificial lights there is a natural airglow in the atmosphere. Even worse, the air above the telescope is always in motion, and this

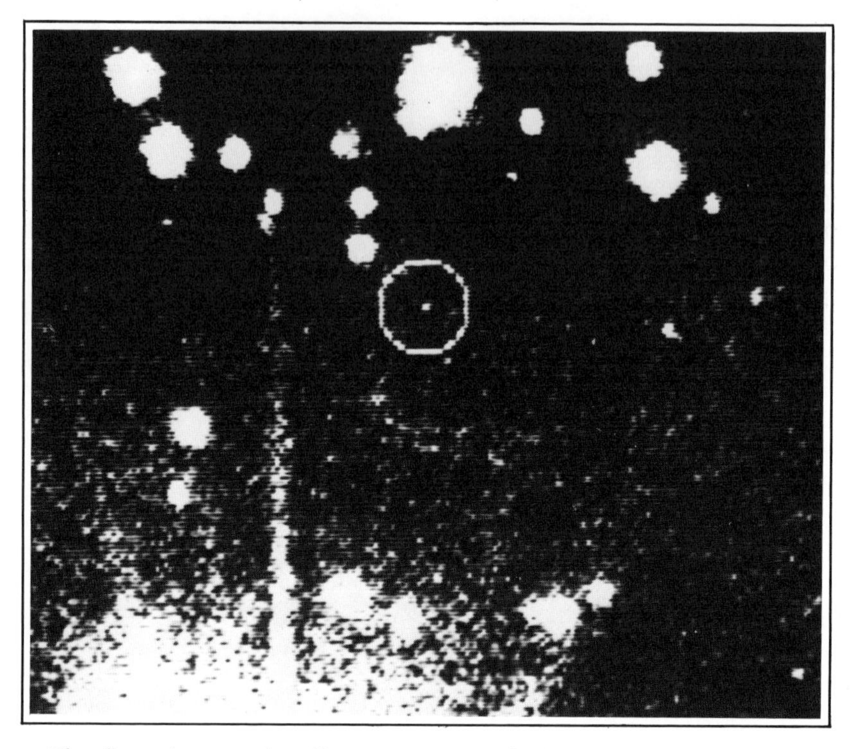

The first picture of Halley's comet (circled) on its present return to the inner solar system was obtained by a CCD camera on the 5-metre Hale Telescope at Palomar Observatory. The CCD's great sensitivity meant that it could pick up the comet when it was 10 times too faint to register on photographic plates

turbulence distorts the images "seen" by the telescope. The smallest details revealed by any large telescope are blurred out to a size of about 1 arcsecond, over 10 times larger than the limit set by the telescope itself. Astronomers can gain some idea of the true size of individual small bright objects by a technique called speckle interferometry; but the only way to achieve a really sharp view of the Universe, with completely dark skies, is to get above the atmosphere.

Hence optical astronomers are looking to satellite-borne telescopes to complement – not to replace – ground-based observatories. Some are specialised, like the Hipparcos satellite for measuring star positions. But most exciting will be the general-purpose Space Telescope. Although it is almost twentieth in the league so far as size is concerned, this 2.4-metre telescope will have unrivalled views of the sky. The Space Telescope will operate as an observatory in orbit, with a lifetime of around 20 years, and it will undoubtedly be the most important single astronomy satellite for the rest of the 20th century. Its cost of £500 million reflects the value that astronomers put on optical astronomy from space. The Space Telescope will graft the technology of the new revolution in astronomy onto the deep roots of optical astronomy, to open keener eyes on the Universe beyond our solar system, beyond our Galaxy and beyond the most distant quasars now known.

30

Precision control at the Anglo-Australian Telescope

RODERICK REDMAN

6 May 1971

Since its opening in 1974 the 3.9-metre telescope in Australia has been involved in most of the important discoveries in the optical (and, more recently, near-infrared) study of the southern sky. Although it comes fifth in the league of telescope sizes, the telescope's performance is second to none, and its computer control is generally reckoned to be the best of any telescope in the world.

A joint project for building a large telescope, to be shared by Australia and Britain, was first suggested by the Astronomer Royal (Sir Richard Woolley) and Hermann Bondi, about 1960. Australian and British astronomers warmly welcomed the idea but negotiations proved protracted and not until April 1967 was agreement reached. The project has been under the joint control of the Department of Education and Science in Australia, and the Science Research Council in the UK, through a Joint Project Committee meeting twice a year. The committee has now been transformed into a more permanent body, the Anglo-Australian Telescope Board.

Following strong advice from the Australians, the telescope is sited on Siding Spring Mountain, where the Mount Stromlo Observatory (itself located just outside Canberra) has had an out-station for some years. Siding Spring is at latitude 31° S, about 500 kilometres due north of Canberra, about 250 kilometres northeast of the Parkes radio dish, and 150 kilometres south of Narrabri, where are to be found the stellar interferometer for measuring star diameters and the circular array of radio dishes for research on the active solar corona. The mountain top is 1200 metres above sea level, one of the higher of an irregular group of hills standing on an extensive and relatively featureless plain, itself about 600 metres

The 3.9-metre Anglo-Australian Telescope at Siding Spring, New South Wales, has the world's most sophisticated computer control system. The mesh cage (centre) protects instruments attached to the Cassegrain focus, and the telescope itself is supported on a massive horseshoe bearing, a form of equatorial mounting

above sea level. The nearest town is Coonabarabran, population 2000, 15 kilometres east as the crow flies. Coonabarabran has reasonably good transport connections by air, rail and road, and there is a good road up the mountain. Over the adjacent countryside there is a thin scattering of other small communities, typically 30–50 kilometres apart, with a somewhat larger town, Dubbo, about 130 kilometres away. There is thus little or no artificial light in the night sky. The sky transparency is generally good to very good, although high-level dust may be blown in from central and northern Australia during the summer. Rainfall is erratic. Coonabarabran records covering over 80 years give annual totals between 330 and 1600 millimetres, with an average of 700. Approximately 1500 hours of "photometric" (steadily clear) night sky are expected annually, with additional hours of poorer quality. The site is rather windy; the average wind speed 15 metres above the mountain top is about 30 kilometres per hour at night. On the basis of experience with the 1-metre telescope which has been in use there for several years, the seeing is reported to be on the whole good, and sometimes very good.

It was a condition of Australian participation in the scheme that the telescope "should follow the Kitt Peak design", the design of the 3.9-metre telescope at that time planned for the Kitt Peak Observatory in Arizona. Although in outline this design has indeed been followed, further consideration has led to a great many important modifications of detail; but this is also true of the Kitt Peak telescope itself, which has departed substantially from the plans that existed in 1967. The AAT is on an equatorial mounting (axes parallel and perpendicular to the Earth's axis), but it should be remarked that possibly the AAT and other telescopes now under construction will be among the last of the big equatorials. The manifest engineering advantages of altazimuth mountings (vertical and horizontal axes) at apertures greater than, say, 2.5 metres, successful experience with altazimuth radio dishes (although optical telescopes have more exacting requirements), and above all great advances in electronic and computer control are together beginning to outweigh the disadvantages of the altazimuth design.

The AAT will have four different focal points, from which an astronomer can choose (Figure 1). The primary focus works at f/3.3. Secondary mirrors give two Cassegrain foci (just behind the main mirror) at f/8 and f/15, and a coudé focus at f/35 in a room adjoining the telescope. (The focal ratios here give the effective length of the combination of mirrors relative to the diameter of the main mirror, so the f/15 focus has an effective focal length of

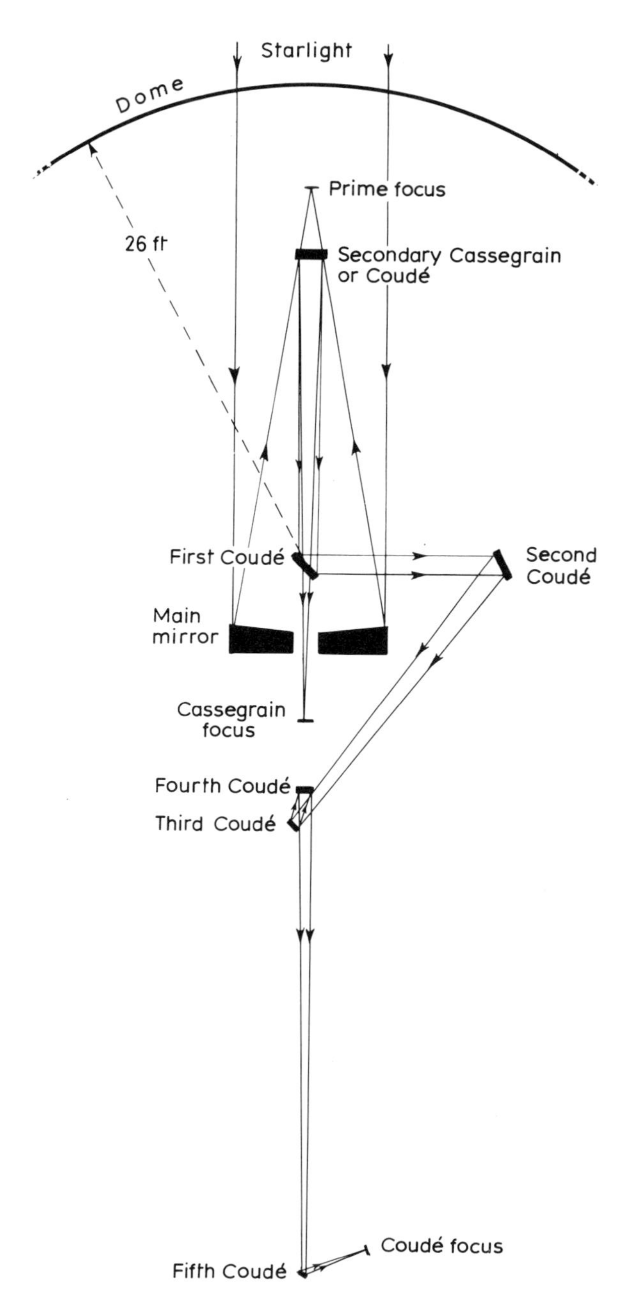

Figure 1 *A conventional reflecting telescope like the Anglo-Australian Telescope has a number of focus positions. The prime focus is at the focal point of the mirror, at the top of the tube. The others are produced by interposing a secondary mirror below the prime focus. The Cassegrain focus lies immediately behind a central hole in the main mirror; the coudé focus, in a separate room, requires five additional mirrors*

58 metres.) The figure of the primary mirror is determined by the requirement that it forms a part of the f/8 focus as a Ritchey–Chrétien combination. In this, the concave primary mirror and convex secondary are hyperboloids whose eccentricities are chosen so that together they annul the optical aberration called first-order coma. This gives a wider field of view with sharp images than the classical parabola–hyperbola Cassegrain system.

Because the primary mirror has a Ritchey-Chrétien figure, some form of corrector must be introduced near the prime focus if the star images there are to be of good quality. Two corrector lenses and also a specially shaped (aspherical) corrector plate will be provided. These cover various fields up to 1 degree in diameter (about 22 centimetres).

Changes between one arrangement and another are by means of three interchangeable top ends for the telescope tube. One carries the prime focus observer's cage, the second has the f/8 secondary mirror, and the third carries the f/15 and f/35 secondary mirrors held back to back, interchangeable with each other by a simple rotation. Provision is being made for easy and accurate checking of the optical alignment of the mirrors. There will be a Cassegrain observer's cage permanently attached to the telescope tube behind the main mirror cell.

The telescope mirrors are made of Cervit, a recently invented vitreous ceramic which has a very low coefficient of thermal expansion (less than 1.5×10^{-7} per degree C, as compared to fused silica 4, Pyrex 30, plate glass 90×10^{-7}). The mirrors should therefore be virtually immune to distortions caused by temperature changes which have plagued large telescopes in the past. The blanks were made by Owens Illinois of Toledo, Ohio, and figured by Grubb Parsons Ltd of Newcastle upon Tyne. Since focus changes will be due almost entirely to thermal changes in the steel framework, an automatic focus control is being designed. Together with automatic photoelectric guiding, also being provided, not only should the observer's work be made easier but, more importantly, high-quality photographs should be obtained more consistently, especially when exposures are long.

The mirrors are all plain discs, with no cellular or ribbed structure. As always, elaborate precautions have to be taken to minimise distortions under their own weight. The primary mirror is supported at the back by 36 pads, themselves supported by air pressure controlled according to the telescope's tilt. The side supports are 24 mechanically counterbalanced pads cemented to the mirror periphery so that they can either push or pull, again according to mirror orientation. The secondary mirrors work mostly face down; each is held in its cell, located against three fixed points, by means of a partial vacuum at the back, again controlled according to orientation. The edge support of each secondary consists of a thin, mercury-filled, neoprene belt.

The telescope is computer-controlled and partly for this reason the motion in both coordinates is not by the traditional worm and wheel, but by spur gears, which today can be made to the required high accuracy. Further, instead of having alternate gears with various motors and clutches, two torque motors on each axis, used in an arrangement which takes up the gear backlash, drive the telescope at all speeds, from fast slewing to slow and highly accurate tracking. A design requirement is that the telescope "setting error" should not exceed 10 arcseconds, meaning that the observer should be able to bring the image of any required star to within this distance of the telescope axis, without further reference to the sky. (The star must of course have known coordinates of sufficient accuracy, which is by no means always the case.) The computer provides the necessary corrections for various effects which will make the star's apparent position differ from that shown in a catalogue: these include precession, nutation, aberration, refraction and mechanical flexure. It then sets the telescope according to the sidereal time, using encoder readings which give the telescope's orientation about each axis. Thereafter the computer controls the ordinary diurnal drive to track the star as the Earth rotates, incorporating varying corrections as appropriate. The photoelectric guider, also supervised by the computer, should then be able to acquire and lock onto the programme star, or alternatively onto some assigned guide star nearby. The aim is to guide to an accuracy of 0.1 arcsecond whenever the seeing allows it.

All this is straightforward in principle, but the computer is not particularly cheap, for it has to do a good many other things too. Furthermore, the software has to be specially developed, mostly in house by astronomers and other staff members, and grow for many years. Some of the control procedures can be complemented by a television display of the star field at which the telescope is pointing, obtained from an auxiliary 60-centimetre telescope beam – the display to be available on the telescope or anywhere in the dome as may be convenient. These techniques are new in optical astronomy and still need a good deal of development work, but the opinion is growing among astronomers that before long it should be unnecessary for the observer to remain on or near the telescope when observations are following some well-defined routine, although doubtless his presence at other times will continue to be required to nurse equipment, or for more experimental investigations. The hope is that long fatiguing hours spent in the dark in chilly discomfort, looking through an eyepiece simply in order to guide the telescope and act as a general monitor, may eventually become a thing of the past. With regard to the computer itself, a steadily growing demand is expected for on-line recording and reduction of instrumental outputs, so that the computer installation must be one that can grow with the years and with the accumulation of technical know-how.

The telescope is mounted with its centre of motion 30 metres above ground level, on a concrete cylinder which stands independent of the surrounding building and dome. The rotating dome, of diameter 36 metres, is of fairly conventional construction, with a steel frame and double roof, up-and-over shutters, windscreens for the shutter opening, and ventilation which can be varied over a wide range, from zero upwards. Owing to the expected windy conditions the dome aperture is smaller than usual in relation to the telescope aperture. Dome rotation and windscreen adjustment then require correspondingly more frequent attention and are controlled by the computer.

The telescope is provided with "common user equipment". For direct photography at both prime and Cassegrain foci there must be corrector lenses, filters, plateholders and the photoelectric guiders. For the Cassegrain focus there are wide-band photoelectric photometers, a fast image tube spectrograph, and another spectrograph of medium dispersion, usable either photographically or with an image tube. There is a large spectrograph at the coudé focus. In addition, active discussions are in progress in both Australia and Britain on a considerable variety of other ancillary apparatus. Approved investigators will, of course, be encouraged to design and construct their own special equipment for one or other focus, provided that they can make a good case for use of the telescope.

The AAT can investigate in depth objects in the southern sky which are not accessible to the large telescopes in the northern hemisphere. Important objects which have been clamouring for closer attention for many years now are the nearest galaxies, the Magellanic Clouds; the southern Milky Way, especially near the galactic centre; the globular clusters; the bright OB stars; and all extragalactic objects south of about 30° S. To these one may now add southern radio and X-ray sources. For some problems, for example extragalactic objects, the southern sky offers an almost virgin field. Very rapid developments of image tubes and television techniques, which are still continuing, are opening up research on vast numbers of faint objects hitherto quite inaccessible. This is of special importance for radio sources, which are generally very faint in optical wavelengths and observable only with great difficulty. The prospect of cooperation with the very successful Australian radio astronomy groups has been one of the great incentives for building this telescope, and it has been kept in view while discussing the already mentioned plans for spectrographs and other equipment on the AAT.

By arrangement with the Australian National University (the owners of the Siding Spring site), the British Science Research Council is erecting a 1.2-metre Schmidt telescope near the AAT. This is to be closely similar to the 1.2-metre Schmidt at Palomar, which has proved such a valuable tool for survey work in the northern sky. Its first job will be to complete the well-known Palomar Sky Atlas to the south pole and thereafter one of its chief uses will be to scout for the large telescope.

31

What's wrong with the largest telescope?

"MONITOR"

18 June 1981

The Soviet 6-metre telescope has never produced results to match its size and its cost – reputed to be some £500 million. The main problem, it seems, lies in the huge mirror itself.

A candid explanation for the continuing poor performance of the world's largest optical telescope, the Soviet 6-metre reflector, is given by astronomer L. I. Snezhko in *Soviet Astronomy* (vol. 24, p. 498). The problem lies with thermal distortions of the main Pyrex mirror, resulting from night-to-night temperature changes of several degrees in the outside atmosphere. Similar temperature changes are experienced on Palomar Mountain, but the 5-metre mirror there is much thinner and comes to equilibrium in a few hours. By contrast, the thick Soviet 6-metre mirror has such a high thermal inertia that it can take up to three days to respond fully to outside temperature changes. Therefore, except on the rare occasions when atmospheric temperatures are similar for many nights running, the 6-metre mirror is always out of shape.

The Russians made frequent and futile attempts to adjust the mirror's temperature during the day to the expected night temperature. They can never succeed because of the disc's immense thermal inertia. The only solution is to replace the Pyrex mirror with a new one of zero-expansion ceramic – a dismal conclusion after 15 years of conscientious planning and construction of the show piece.

32

Schmidt telescopes widen the field

PETER FELLGETT

10 July 1969

A Schmidt telescope is a hybrid of reflector and refractor, which can photograph very wide regions of sky on special plates up to 35 centimetres square. It produces far too much information for traditional methods to handle, and the necessary advances suggested in this prescient article have now become reality.

One of the most far-reaching events in the history of astronomical observation took place in the early 1930s at the Hamburg-Bergedorf Observatory when Bernhard Schmidt conceived the telescope system which bears his name. He constructed the first example with his own hand; the left hand to be precise, because he had lost the other one in a too successful boyhood attempt to make explosives. The prototype was of 44-centimetres aperture, and worked at a focal ratio of f/1.75 over a field 16 degrees in diameter. These values seem startling even today.

The original concept of an astronomical telescope, from Lippershey and Galileo in the 17th century, has been described as "a large lens at one end and a small lens at the other". The observer looked through the small lens, and his astronomical results were what he took home in his notebook. This approach had indeed been modified, around the beginning of the present century, by the introduction of photography and the consequent development of "astrographic" field-imaging telescopes; and by the rise of astrophysics, associated with the introduction of spectrography into the observational techniques of astronomy. Nevertheless, in 1918, the effective date of commissioning of the 2.5-metre Mount Wilson reflector and also the effective epoch of the optical design concept of the 5-metre Palomar telescope, the "big end/little end" concept was evidently still dominant.

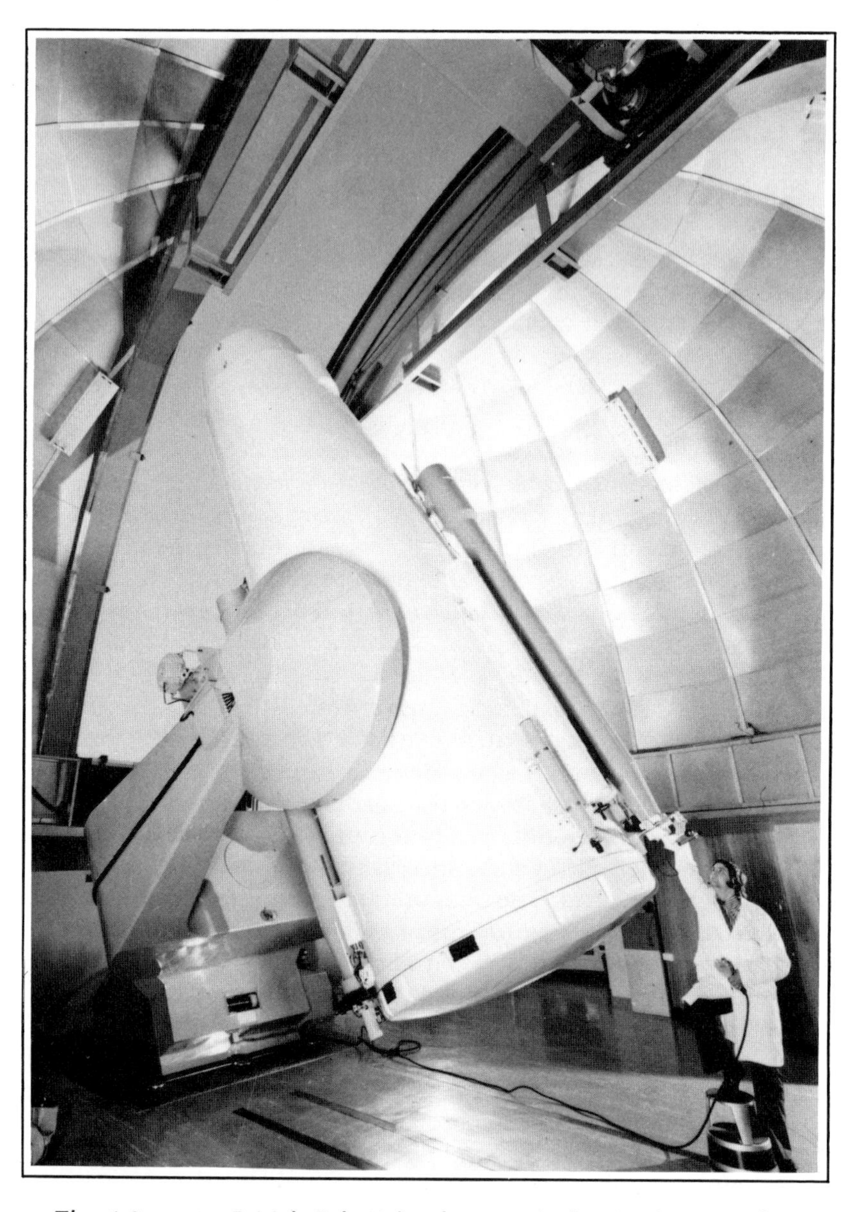

The 1.2-metre British Schmidt telescope is the Anglo-Australian Telescope's neighbour at Siding Spring. Operated by visiting astronomers from the Royal Observatory, Edinburgh, this purely photographic instrument has recently completed a thorough survey of the southern sky, pinpointing objects for its larger neighbour to examine in more detail

The significance of Schmidt's invention lies in the enormous increase it provided in the power of the astronomer to record information about the sky. A modern f/3.5 Schmidt, working over a field 5 degrees across, produces a spread of each star image (due to aberration) not exceeding 0.4 arcsecond in diameter, and this spread is symmetrical. By comparison, a telescope like the 2.5- and 5-metre working at the same focal ratio gives a field less than 1 arcminute across for a maximum coma spread of 0.4 arcseconds. The currently fashionable Ritchey–Chrétien gives 20–30 arcminutes at f/8, or up to 60 arcminutes by the use of field-widening lenses. Even the latter figure represents only 4 per cent of the field area of the f/3.5 Schmidt.

In basic construction, a Schmidt telescope has a large curved mirror at the bottom of the tube, with a shape which is part of a sphere ("spherical") rather than parobolic or hyperbolic. The top end is closed by an almost flat glass corrector plate, whose exact shape is calculated to correct for aberrations caused by the spherical primary.

A Schmidt thus has three significant optical surfaces, two in the corrector plate, and the mirror surface itself. These need to be held in the correct relative positions with tolerances of better than 10 micrometres for some of the adjustments. However, there are rather easy optical tests for the principal adjustments, which can indeed be exploited for servocontrol. Moreover, the whole process is greatly helped by the fact that a sphere has no unique axis, consequently several of the degrees of freedom of maladjustment can be ignored; correct adjustment is, of course, just as important for testing during manufacture as in actual use. With care and skill in figuring and in "tuning-up" on site, it is possible to exploit these favourable factors so as to achieve, in practice, substantially the full theoretical performance of a Schmidt. Of the telescope systems which we currently know how to make and to maintain in adjustment, the Schmidt can collect information at a rate an order of magnitude greater than that of any other system.

The exploitation of this power requires a revolution, which is not yet complete, in astronomical thinking. It is also dependent on two non-astronomical developments of the past quarter-century: automatic control and digital computers.

A Schmidt of quite moderate size can record photographically as many as 50 000–100 000 stars in a single exposure of a few minutes' duration. Moreover the very precise imaging of the Schmidt means that these star images can be measured with great accuracy for both position and brightness. It is clearly impracticable to make these measurements by hand and eye at anything approaching the rate required to keep up with the enormous output of a Schmidt. The necessary measurements can be made using modern developments in digital distance measurement and photoelectric photometry, and in servocontrol and electronic methods. A number of projects for automatic measuring machines of this kind are current.

Traditional hand methods of reduction and interpretation of the data so produced are equally inadequate. Fortunately the work is almost ideally suited to modern digital electronic computers, but changed attitudes are needed on the part of the astronomer if this powerful facility is to be fully exploited. The telescope and computer must each be regarded as subsystems of the overall observation system, the automatic measuring engine being a further subsystem which interfaces the telescope to the computer; no one subsystem can properly be designed in isolation without regard to the optimisation of the observational system as a whole. The data stored in the computer files will be "reduced" in the conventional sense of having been expressed in magnitudes, colour indices, right ascension, declination, and so on, by reference to conventional standard stars. However, they will be regarded as raw data by the astronomer, who interrogates the data bank by program in order to extract the significant information of interest to him. He must obviously plan his approach so as to avoid being overwhelmed by too copious output; for example, he can ask for detailed information about stars having specified rare characteristics, or else ask for statistical information about common types of stars.

33

Counting the stars by computer

ED KIBBLEWHITE

18 August 1983

Lasers and computer control have led to a machine which dramatically speeds up astronomers' analyses of photographs and has already turned up new kinds of astronomical objects.

With a modern wide-angle telescope, an astronomer can take a photograph of the sky that records over a million images of stars, galaxies and quasars, in an exposure of about an hour. But he may then have to spend weeks measuring the photograph by eye with a hand-operated measuring machine fitted with a microscope. And the work involved means that he will usually look only for the kind of object that he is interested in. The rest of the information on the plate – including, possibly, unsuspected types of astronomical objects – goes to waste.

Astronomers observing at other wavelengths – from X-rays to radio waves – now routinely use computers to assist their analysis, and the time is clearly ripe for optical astronomers to follow suit. But the sheer quantity of data on an astronomical photograph makes this a daunting task. William Bond at Harvard recorded the first photographic image of a star, Vega, on a polished silver plate in 1850. Since then, sustained improvements in the design of telescopes and in photographic technology have meant that, until very recently, the rate of obtaining information on photographs has doubled every seven years, until a single exposure shows over a million images – far more than the total number of sources known in the whole sky at any other wavelength. The task of measuring these images automatically presents a major challenge, both in instrument design and in computer analysis.

The first automated machine for analysing photographs was developed in the IBM Watson Research Laboratory in the late 1950s, and was one of the earliest numerically controlled machines of any type – one of the first, that is, to be controlled by instructions from a computer. An astronomer measured the positions of the required stars approximately, and the machine went to these locations and automatically centred on the images to give the precise positions. The latest types of automated machines are much more ambitious. They convert the whole photographic plate into a grid of numbers and use fast computers to analyse the data, measuring more objects in an hour than one man could measure in an entire year's work.

To understand why these machines are necessary we have to study the photographs themselves. A cursory look at a photograph shows a number of bright stars that appear as black dots in the middle of a cross. The stars are black because we look at the photographic negatives; the brighter stars appear bigger because their light is scattered in the photographic emulsion; and the cross is caused by diffraction of light by struts in the telescope. Looking more closely, we see enormous numbers of small dots. The sharper dots are faint stars, and they can tell us about the large-scale structure of our own Galaxy, the Milky Way. The fuzzy dots are distant galaxies. The original photograph records a million faint galaxies, many so distant that it has taken 4000 million years for their light to reach us. Over this period galaxies slowly change and interact with one other, so by studying these objects we can ascertain the large-scale structure of the Universe and see what it was like 4 billion years ago. There are also a few, more exotic objects like quasars. These look like stars, but are in fact the brightest and most distant objects known in the Universe. They are very rare and difficult to find, with perhaps only a few hundred quasars among a million objects on a photograph. Indeed it is almost certain that the photographs show images of even more interesting types of object still undiscovered, and one of the main aims of building automatic machines is to find such objects.

There are three main problems in analysing this data automatically. Each plate contains a lot of data – some 2000 million individual picture points. We are interested in the fine detail of these images, so we must measure their properties very accurately. Finally, we must be able to select out the data of interest from the vast amount of information produced from each plate. All three problems have to be solved together if we are to produce effective machines. I certainly had no idea of the problems when I decided, in 1966, to try to build the fastest machine in the world for measuring these photographs as a PhD project.

At the time, the first modern measuring machine was being built at the Royal Observatory, Edinburgh. Designed by Peter Fellgett, it was called GALAXY and it could measure accurate positions of 900 stars per hour. GALAXY was limited by using the relatively faint spot of light from a cathode ray tube to scan the plate, and by the inertia of the heavy plate-carrying table which had to be repositioned for every star measured.

I intended to use a bright laser beam to scan the plate, to overcome the rather dim light output of a cathode ray tube scanner. A pair of mirrors would deflect the laser beam over one square centimetre of plate, and the heavy table would only be used to move new areas under the scanner as required. A high-speed laser beam deflector would scan the plate at low resolution to pick out the individual images. In 1968, however, laser-scanning technology was very new and, while mirror reflectors proved easy to build, a suitable high-speed scanner had not yet been devised. I tried various approaches: setting up sound waves in liquids of high refractive index – liquids that were generally very poisonous; and building mirrors on magnetic deflectors, which became incandescent at the power levels needed to operate them. In total the workshop made 23 prototype scanners before we developed a simple high-speed scanner using mirrors driven by piezoelectric transducers. The prototype system did not work very well (especially when shown to visitors), but after six years and a PhD I did at least know the problems. The Science Research Council (now SERC) provided funds to build a fully working machine. The Council showed great faith in someone who held one of the lowest posts in the university!

Throughout the whole development, the SERC has been enormously helpful and supportive. We split the project into two. The design and construction of the scanning system – the microdensitometer – was handled by SERC's Cranfield Unit for Precision Engineering, following up our experiences with the prototype. Here at the Institute of Astronomy at Cambridge we set about developing the techniques for handling the vast amount of data which would result. The result is the automated photographic measuring machine.

In the final design of the microdensitometer, the photographic plate is held on a massive moving table. Although the table weighs 5 tonnes, it can be located to a precision better than a micrometre. Its great mass means that it is less susceptible to temperature fluctuations: change of 0.1 degree C in the ambient temperature produces errors of a third of a micrometre in measured positions and for some

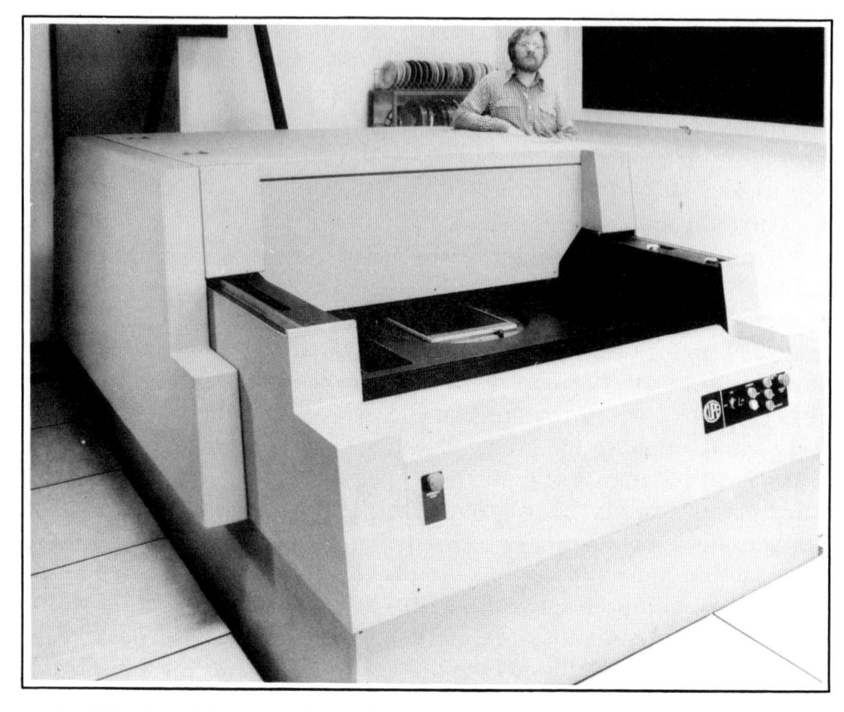

Ed Kibblewhite stands by his automated photographic measuring machine

projects we need to measure the relative positions of stars to better than a tenth of a micrometre. Despite the table's mass, powerful motors can position it in less than a second. The scanner uses a 5 mW helium-neon laser as the light source. This beam is expanded to a diameter of 10 millimetres and passed through an acousto-optic deflector. The deflector sets up high frequency sound waves (100 MHz) in a crystal of lead molybdate, and the wave pattern forms a diffraction grating which deflects the light beam through an angle dependent on the frequency of the sound wave. By altering the sound wave frequency, the beam is scanned across the plate. The beam is focused to a spot 8 micrometres in diameter on the emulsion, and each scan is 2 millimetres long. One of the big advantages of the laser scanner is that its intensity and position can be maintained very precisely, and our microdensitometer is probably the most accurate in the world.

The plate is moved under the laser beam and is thus scanned like a very high resolution television picture. The light passing through the photograph is collected by a photomultiplier tube under the table. The signal is amplified and the brightness converted to digital form, on a scale of 1 to 4096. The system typically samples a quarter of a million spot-sized areas every second. The microdensitometer can reach much higher speeds, but the system is currently limited by the speed with which the computer can process the data.

The computer system evolved through a number of stages. In the 1960s, the data handling problems were fairly new. We started with an analogue system which would analyse circular scans around each image on the plate, but its development was uphill work. Fortunately, in 1967 I had met by chance Jim Tucker, who was trying to analyse cells of the cervix of the womb to detect cancer. This involved measuring the sizes of the cell nuclei which were stained black, and looked just like star images. We scanned small areas of photographic plate under a television camera, and processed it with Tucker's software. This was probably the first time astronomical images had been studied in this way and, unlike the other approach, was immediately successful. The data were accurate, and we could also take account of the changing brightness of the background "fog" over the plate. All the early machines had measured only the position and brightness of the images of stars, and for these small, sharp images the background is unimportant. But we wanted to measure the images of galaxies as well. Faint galaxies look rather like stars but are "fuzzy", that is, they have a faint extended halo which contains much of their light. To find the halo brightness, we must accurately subtract the sky background. Tucker's approach was very attractive, because he had had to solve the problem of variable background to measure his cell nuclei.

The only problem was that even the biggest computers were much too slow to analyse the data in a reasonable time. Although it was considered unwise to base the whole project on a new method, I had been given some money in an early design-study grant for a purpose-built "pipeline" computer. Many of the mathematical operations needed could be done in sequence, so that we could build a number of separate modules, each one doing one function and then passing the data on to the next, so forming a pipeline. The computer became rapidly more complex as the team – Mike Bridgeland, Tony Hooley, Dave Horne and myself – wrote simulation software to test the system, and we designed it in innumerable "project meetings" – many held late at night in the Panton Arms. The final system was to consist of three computers linked together; it would be at least 10 times faster than the main university computer, an IBM 360/195;

and it would be sufficiently flexible to handle the foreseeable needs of astronomers. The new scheme looked so good that we decided to abandon the original design and go for this faster computer. With only a fixed amount of money, however, we had to build the computer with our own hands. Although it was very hard work, it did give us flexibility in the design, and was probably a good thing. In this we followed the tradition of the universities. When you contract out work to industry it is crucial to change the specification as little as possible, because these changes are usually costed in a different way from the original specification and work out very expensive (and are indeed how the firm makes its profit!). The machine thus does not benefit from experience gained during its construction. We had maximum design flexibility, and consequently a better computer – but our approach meant losing several years in building the machine.

The computer of the automated photographic measuring machine has many tasks to perform. The first problem is that the darkness of the image on a photographic emulsion is not directly proportional to an object's brightness. During the original exposure, however, the astronomer will have put a set of calibration squares of light intensities from an artificial source on the corner of the photograph. The computer uses these squares to convert the measured transmission of the plate at each spot into a brightness value. We always use a very small spot of light to measure the plate transmission. If we want to produce the effect of a larger scanning spot, we use special purpose hardware to combine the brightnesses of the small picture elements – a task which would take a long time on a commercial computer.

To understand the rest of the system, let us look at how the machine works in practice. Once the plate is put onto the table, the operator types in the position in the sky of the centre of the plate, and the system uses a star catalogue to line up the plate automatically to celestial coordinates. The plate is then scanned to measure the background. This is done by a special background processor, which investigates a quarter of a square millimetre of the plate at a time. In each small area (consisting of 64×64 samples) it finds the number of times every possible level of intensity occurs. Since most of each area consists of background sky, the background level is the most commonly occurring intensity level. We must be able to measure stars within nearby galaxies, which have backgrounds changing markedly with position, so the system determines the background for every sample we measure. After the plate has been scanned, another hardware device interpolates the background between the measured points spaced at half-millimetre intervals to produce a smoothed map telling us the background at any point on the plate.

We now scan the plate again. A threshold is set just above the background level, and all points below the threshold are rejected. Because the images of galaxies must be measured very near the background level, random fluctuations in the density of grains in the emulsion produce many false "noise images" above the threshold – often 100 million over the plate. These fluctuations are, however, smaller than genuine images, and they are removed by an electronic filter which looks round every image and takes out those which are too small. This filter reduces the number of noise images present in the data by over 90 per cent, without affecting any real images. The data is then sent to a slice processor. The processor looks at the data, line by line, and computes the position and intensity profile in one dimension of every part of an image above background, while another processor works out which slice on each line belongs to the same image as slices on other lines. This associative processor provides input for the third and final piece of hardware, a high-speed programmable computer which combines all the data for every image to give its position, total intensity and intensity profile in two dimensions. The computer extracts almost all the information on the plate, and we can also program it to carry out other specific tasks. The data is stored on magnetic tape, for processing later off-line on a big computer.

We are fortunate in having a powerful VAX computer in Cambridge as part of SERC's Starlink network of computers dedicated to analysing astronomical information of all kinds. The VAX can handle the large amounts of data efficiently and at high speed, but nevertheless it takes longer to analyse the data than it does to scan the plate. Every researcher wants to extract different kinds of information, and a library of special software is slowly building up to assist them.

Having built the machine, we now run it for the SERC as a national facility, and it is used by astronomers from all over the world. A large fraction of our time is spent helping users; often they need new features and the machine is constantly developing. In particular, we find we need to analyse in more detail objects found in the first scan of the photograph. To give a flavour of the work now carried out on the machine I will describe three current projects.

One of the key problems in the study of galaxies is the problem of "missing mass". We believe that galaxies may be more massive than they appear from the starlight we can see, and the only way to measure a galaxy's mass is by studying the orbits of objects a long way out from its centre. Globular clusters are ideal objects for this work. They are dense collections of stars which are intrinsically bright and orbit a galaxy out to a distance of about a third of a million light years. The nearest galaxy of interest for this work is the Andromeda Galaxy, but it has only a handful of globular clusters in its outer parts. Among the millions of other objects on the photo-

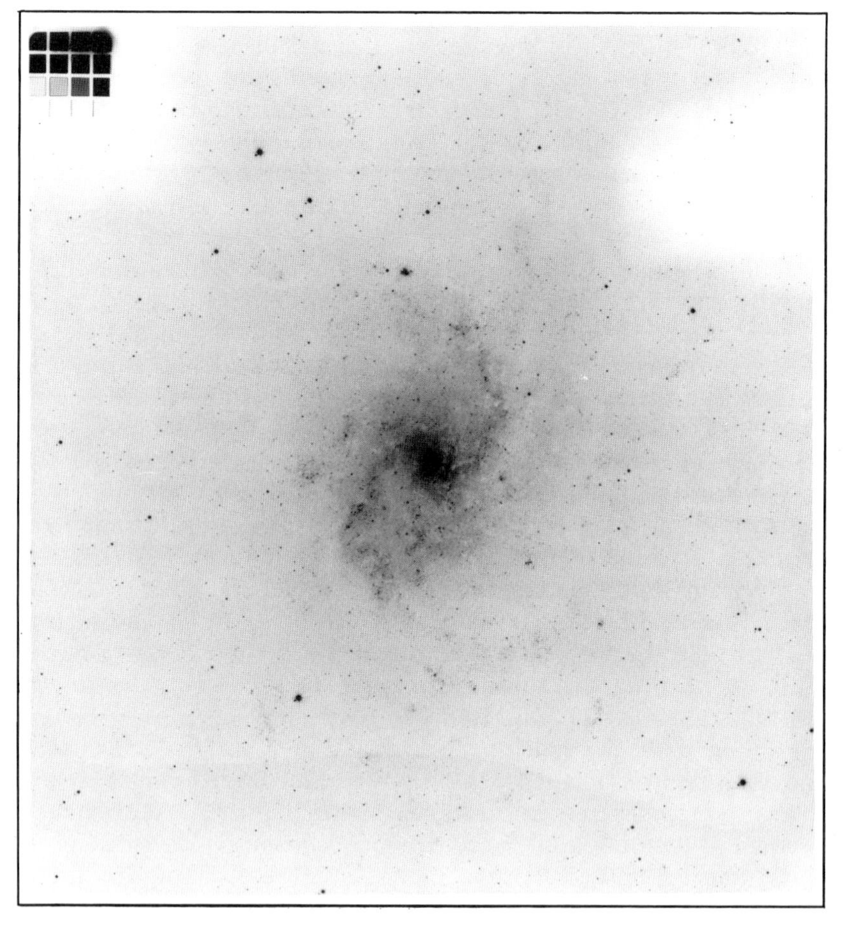

The APM can scan photographs of the galaxy M33 (above) to plot distribution of blue stars (right)

graphic plates, even the most careful workers cannot distinguish by eye these few clusters from distant galaxies on wide-area photographs. The machine, however, can detect very subtle variations in the shape of these objects and can separate them from almost all the foreground stars (in our Galaxy) and from the galaxies beyond. More important, we can measure the brightness of all the objects on a number of photographs taken at different wavelengths, and can collate all the data on the computer to measure the colours and temperatures of each object. This enables us to reject images of distant galaxies. We are currently searching an area of a 100 square

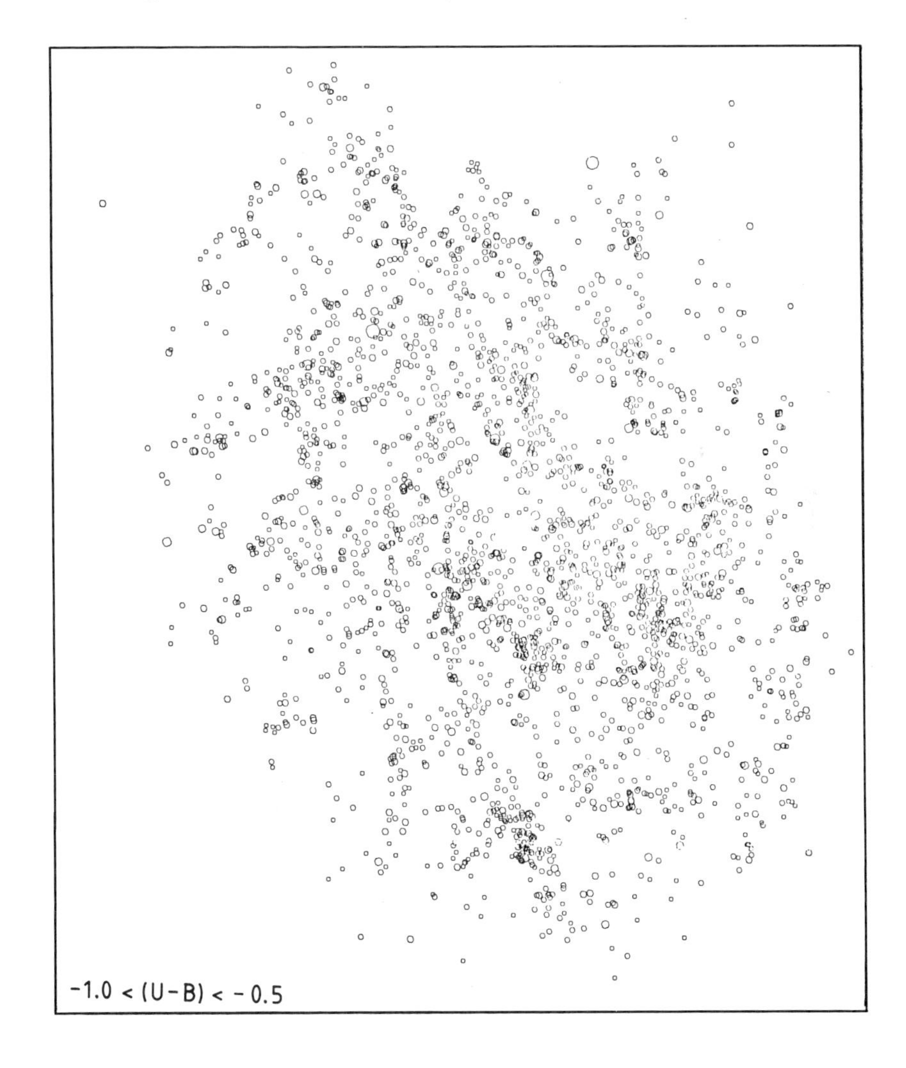

degrees around the Andromeda Galaxy to find the few very distant globular clusters. These will then be studied on large telescopes and the mass of the galaxy obtained.

Being able to measure the colours of stars in nearby galaxies is crucial for many other problems. Wendy Freeman is picking out the blue-coloured very hot stars which are bright, massive and have very short lifetimes of a few million years or less. Because they do not have time to move appreciably over their lifetime we can see where the stars are born.

In a nearby spiral called M33, the positions of young, blue stars have been found on our machine. They are formed mainly in dense clouds within the spiral arms. However we have found some star formation occurring between the arms – an unexpected, and important, observation. The distribution of older stars, which have had time to move from their birthplace, is very different. For the first time we can measure the rate of star formation, and how it depends on the distribution of interstellar gas, for large numbers of nearby galaxies. These measurements provide crucial data for solving problems of the structure and evolution of galaxies.

Perhaps our most challenging long-term project consists of studying small regions of the sky in great detail. The project involves a team of astronomers from the United States and Cambridge, under the direction of Cyril Hazard, who measured the first accurate position of a quasar in 1962. Although that quasar, 3C 273, is a powerful radio source, most quasars are only detectable by their output of light, much of which is in the form of spectral lines. If a large thin prism is placed in front of a telescope, the photograph shows all objects as spectra rather than straightforward images. We can readily pick out quasars by their distinctive emission line spectra. Cyril Hazard spent many years identifying quasars in this way. The trouble with this technique is that the human eye tends to spot quasars at a particular range of distances (corresponding to a redshift of their spectral lines between two and three) more easily than nearer quasars, with a lower redshift, because a redshift of two or three moves the characteristic strong ultraviolet emission line of hydrogen, Lyman-alpha, into the blue part of the spectrum. The eye also tends to miss quasars with weak spectral lines. We are now developing a totally automated detection system which forms an enormous database of objects in the sky, measuring not only prism plates but a number of photographs taken through filters which pass only a narrow range of wavelengths. These measurements allow us to build up the spectrum from the ultraviolet to the infrared, and the data are also being combined with measurements from radio telescopes.

The technique has already turned up a new kind of astronomical object – stars in our Galaxy which emit radio waves. We picked these objects as very weak radio sources from sensitive radio surveys made in collaboration with Jim Condon using the Very Large Array radio telescope in New Mexico. We thought some of the objects would be extremely distant quasars, and we were initially puzzled to discover that the first two objects found on a photographic plate at radio-source positions were actually stars, when their spectra were examined in detail with the large Multiple Mirror Telescope in Arizona. The optical positions, however, coincided exactly with radio sources and we knew we had finally discovered "radio stars". Astronomers have searched unsuccessfully for radio-emitting stars since the dawn of radio astronomy, some 40 years ago. These new stars are slightly fainter than our Sun, and they probably have a strong radio-emitting outer atmosphere (corona) and may be similar to some X-ray stars.

We will continue to extend the capabilities of the system. One of the most challenging areas is to write software so that the computers themselves can recognise and classify common objects and pick out rare and interesting objects for us to study. We will then be able to look at the myriad objects that make up the Universe to the limit of our telescopes.

34

Computers make light work for astronomers

"TECHNOLOGY"
6 March 1980

Radio astronomers have naturally used computer power to analyse and display data, because a computer is an integral part of all modern radio telescopes. Now British optical astronomers have a special system that harnesses computer processing to optical astronomy.

British astronomers now have a computer network to help them to process the vast quantity of data that they collect from telescopes and space satellites. Called Starlink, the £1.8 million network links six computers, and it provides the astronomers with processing and graphics facilities, as well as shared software.

Computers help astronomers in two ways. First, digital techniques developed for a particular telescope reconstruct a two-dimensional "map" – a digital image that shows, crudely speaking, how much radiation comes from a part of the sky, or forms part of the spectrum of an object.

Second, the computer scans the data from the telescope to find the information the astronomer wants about the physical and chemical processes that he is studying. For example, the scientist might want to calculate the amount of light that comes from a source such as a galaxy. He has to decide how much of the region on which to concentrate, and which of the intervening objects, such as stars, to eliminate from the calculations.

Building up images from observations with telescopes involves huge volumes of data. For instance, a scientist collects roughly 8000 million bits of data in one day's observation with an optical telescope. Researchers in this area share with other computer users another problem – the time it takes to write software to handle the data. But now those data are coming from almost all telescopes in a reasonably uniform format, astronomers are wasting effort by writing software to tackle essentially the same problem.

In 1978 the Science Research Council (SRC) set up a panel to plan how to eliminate the duplication of software and cut costs. Headed by Mike Disney, professor of astronomy at the University of Cardiff, the panel initially considered asking astronomers to use a network called the interactive computing facility (ICF) that the SRC already provides. This links university users with computing facilities at the Rutherford Appleton Laboratory in Chilton, Oxfordshire, and at other places around the country. But the ICF could not cope with the volume of data the astronomers would generate, and anyhow would do its job of passing data between different machines too slowly.

The panel finally decided to set up minicomputers to handle image processing at six regional centres. Astronomers who do not have access to one of the computers will have to travel to a centre with their data on magnetic tape. Software packages for performing various types of analysis will be gradually built up to form a central library. So that changes to programs are incorporated uniformly, software will be called up from its "home" and transmitted over telephone links.

The SRC has selected for its computers six machines made by DEC of the USA. Although the SRC wanted to buy the equipment from a British company, the panel found no British firm that could supply suitable systems which operate on a 32-bit word length. This facility was considered essential because the computer has to handle complex data.

The network will be run from the Rutherford Appleton Laboratory. The other five centres are at the Royal Observatory in Edinburgh, the Royal Greenwich Observatory, London's University College, the University of Cambridge and the University of Manchester.

The Starlink project will process images from many types of telescope. Among them are the Anglo-Australian Telescope and the UK Schmidt telescope, which are both at Siding Spring in Australia.

35

Quest for the world's best observatory site

ROS HERMAN

7 June 1979

When British astronomers wanted to find the best possible location for their newest optical telescopes, a thorough testing of possible sites led them to the Spanish island of La Palma, in the Canaries.

Bad weather makes Britain a rotten place for looking at stars. Cloud cover, rain, sleet and snow, air pollution and light from cities all serve to take observation time away from the optical astronomers, and distort the images they do manage to see and photograph through their telescopes. This has not discouraged compulsive star-gazers in back gardens, but it has meant that professionals have often gone winging off to other parts of the world. Today, a quarter of Britain's practical astronomers are observing at or helping to set up facilities in the southern hemisphere.

The reception of radio waves is almost unaffected by weather, so observation of radio signals from outer space can, and does, go on in fair weather or foul, day and night. But with excellent maps of the radio sky for the northern hemisphere now available, thanks to the efforts of radio astronomers in Europe and the United States, there is a need for good optical studies of the objects being investigated at radio frequencies.

Wanted – one site in the northern hemisphere, not too far from the UK, with good weather, clean air, high mountains and a gentle wind flowing smoothly over the area; also a stable and friendly government, if possible.

The quest – officially called the Site Testing Project – began in 1970, and in April 1971 was taken over by Bennet McInnes, then leader of the satellite tracking section at the Royal Observatory, Edinburgh. In late 1971, three teams of enterprising and resourceful young men, mainly students and amateur astronomers, were preparing to spend a year assessing three sites: one in Italy, one at Calar Alto in south east Spain and one at Montana Izana on Tenerife where there is a small Spanish observatory. This was a new way of site testing – in the past it had been done by senior researchers, who had set up automatic recording equipment, visiting it only occasionally to see how it was faring. Bennet McInnes felt that this was not very effective – faulty equipment often produced misleading information, and more reliable impressions and records could be collected by people on site for a long stretch of time.

The site testers slept during the day and got up at tea-time to make sunset observations of cloud cover, air temperature, absorption of starlight and sky brightness. All these were repeated hourly during the night. The object of the exercise was to establish the number of "usable hours" – the number of hours during which certain rigidly defined criteria of weather and visibility were met.

For each usable hour, site testers endeavoured to place a quantative measure on the quality of image that a telescope could form under the prevailing conditions. They used a method pioneered by Merle Walker of the Lick Observatory in the site survey he carried out in California in the 1960s. Observers use a 15-centimetre telescope – the size used by many amateurs – pointed at the Pole Star. A photographic image is formed of the star as it moves across the sky, and the width and quality of the image give a good measure of visibility.

It took eight to nine months to find out that the Italian site did not look good; the people there packed up and went home. By this time relations with Spain had become far from smooth, so the team leader waiting for his team at Calar Alto left the site to the Germans, who had already set up an observing station there. All that remained was the group on Tenerife.

A couple of observers from the group spent six weeks on La Palma, a small fertile island, with a population of 50 000, that had also attracted the attention of a group of European solar observers. There is little air pollution, and the prevailing wind at mountain top level flows smoothly around the island. Another advantage, common to all the sites in the Canaries, is that it is a region of high atmospheric pressure, where for three-quarters of the year there is a temperature inversion in the atmosphere which forms a sort of lid, keeping moisture and dust at low levels, generally below 1500 metres. At these times, the observers, at a height of 2400 metres, enjoyed a clear blue sky above and rarely paused to look down on the clouds a kilometre or so below.

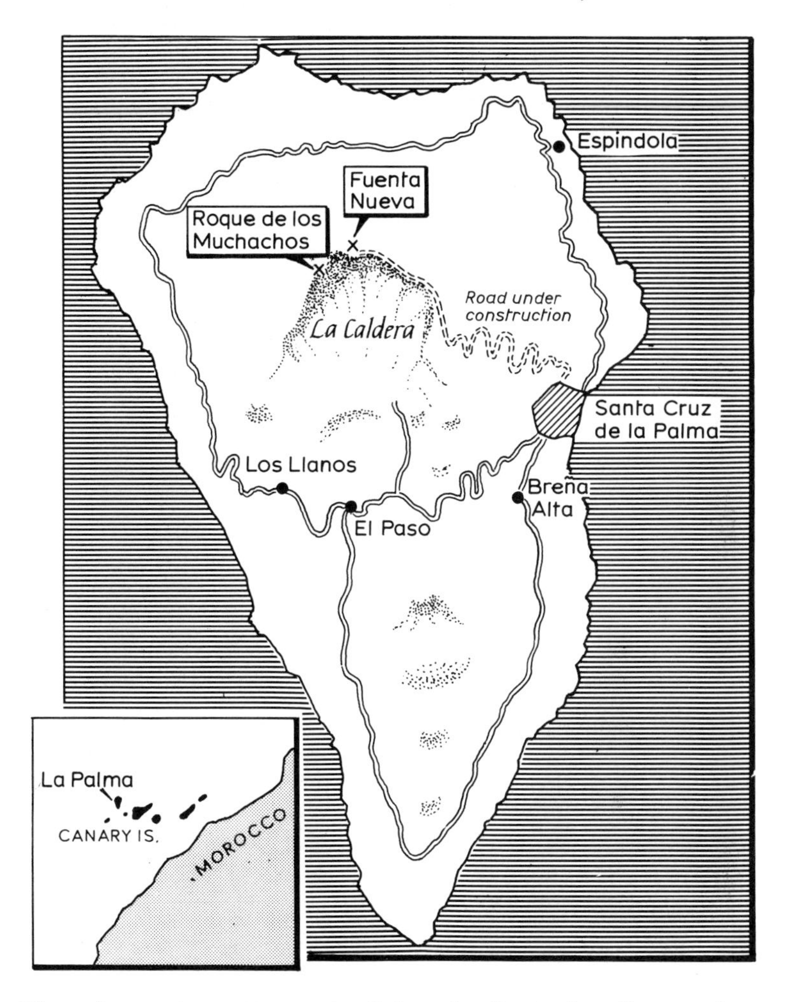

The observatory site on La Palma is situated well away from civilisation – ideal for observing, but difficult for constructors

---

(Opposite) *The telescopes of the new observatory are perched on the rim of La Palma's vast caldera. At left is the stone pier for the original small site-testing telescope; in the centre the 1-metre telescope and to the right the Isaac Newton Telescope*

So, in 1972, the testers reported that, despite the short time they had worked there, they were extremely happy with conditions on La Palma, and were convinced that it was the right place for the observatory. This looked too good to be true, and it was. The Science Research Council called back the last group of researchers, because officials felt the necessary diplomatic negotiations would be too difficult in view of the political situation. The site testers did not share the officials' pessimism, but, like it or not, they were back to square one.

For two years they tried out one site after another, with very little success. They went to Fogo Island (in the Cape Verde Islands) and to Madeira. Having exhausted the possibilities in the Atlantic, they considered both California and Mexico before plumping for the big island of Hawaii, where they found very good observing conditions – but were they good enough to justify the enormous cost of ferrying people and equipment that far? Any fantasies British astronomers might have nursed about the tropical island's way of life were quickly forgotten when at the end of 1974 site testers got permission to go back to the Canaries to work alongside their Spanish colleagues in an international team.

The moves towards democracy that followed the death of Franco in 1975 brought an upsurge of interest among Spanish astronomers, who found that joint projects with other countries were suddenly encouraged. Although the change was certainly for the better from the point of view of the British astronomers, it did cause problems at first. The entire negotiating team changed and British scientists and diplomats had to build up relationships with the new people. The slightly embarrassing confrontation between Britain and Spain was defused by the presence of the two new participants, Denmark and Sweden. But the negotiations took four years – much longer than was originally expected – and the signing of the documents was a great relief to all concerned. The whole site, which will also include telescopes built by Swedish and Danish astronomers, will be known as the Roque de los Muchachos Observatory.

36

British telescopes head La Palma observatory

NIGEL HENBEST

10 November 1983

The United Kingdom is constructing several telescopes on the peak of La Palma, in the Canary Islands, culminating in the great William Herschel Telescope, a reflector with the world's third-largest telescope mirror. The skies here are so clear, and the air so steady, that several European nations are collaborating in setting up an international astronomical observatory on La Palma's summit.

The observatory site is officially Spanish, belonging to the Instituto de Astrofisica de Canarias, and they have spent an estimated £5 million on providing the necessary services – an access road through rough volcanic terrain, an accommodation block and electric power and telephone lines. The Spanish have not yet built telescopes of their own on the peak, but in return for these services, and acting as "host" to the other countries, Spanish astronomers are given one-fifth of the available time on each of the telescopes. Although the telescopes on La Palma belong to the countries that built them, the agreement insists that as well as the allocation of time to the Spanish, 5 per cent must be allocated to international projects involving teams of astronomers from the participating countries.

The first telescopes to be completed were two Swedish instruments. A 60-centimetre reflector began observing in 1982. Its main purpose is to study moderately bright stars through a series of colour filters, each passing only a narrow range of wavelengths. This "narrow-band photometry" was devised by the Swedish astronomer, Bengt Strömgren, and it provides a simple and accurate method for measuring the temperature of a star, the abundance of elements (in the cooler stars) and the proportion of a star's light that is absorbed by dust in interstellar space. The telescope can also measure the brightness of emission lines from stars, and the polarisation of starlight.

The second Swedish instrument is a solar telescope. It is essentially a 21-metre long refracting telescope, fixed vertically in a tower 16 metres high. A flat mirror reflects sunlight into the lens at the top of the telescope tube, which is evacuated to prevent air currents from distorting the image, and the lens forms a crisp image of the Sun in a laboratory below ground level. So far, Swedish astronomers have tested the system using a simple lens only 20 centimetres across, and have already obtained some of the most detailed photographs of the Sun's surface ever taken. They plan to substitute a 45-centimetre lens, made from two types of glass to prevent false-colour fringes, and should then achieve even better results.

The Danish contribution is a telescope designed to measure star positions with very high precision. The Carlsberg Automatic Transit Circle is a refractor with an 18-cm lens, which is limited to swinging up and down the north–south line in the sky, measuring the positions of stars as the Earth's rotation brings them across this line. The University of Copenhagen operated this telescope for many years in Denmark; on La Palma, however, it will be a joint Anglo–Danish operation, as the Royal Greenwich Observatory has computerised the telescope so that it will work entirely automatically. If the transit circle's title looks familiar, there is a reason: the cost of the telescope itself was borne by the science foundation of the famous Danish brewery!

The Carlsberg Automatic Transit Circle is now being tested on La Palma, and should start work by the end of this year. It can measure the positions of 1000 stars every night, and by repeating the observations and averaging the results, the typical errors in a star's position can be reduced to only 0.1 arcsecond – the diameter of a 1p piece at a distance of 40 kilometres. The work is necessary to help in identifying faint radio and X-ray sources, especially as the positions of stars given in existing catalogues will deteriorate with time, because the stars themselves are moving. Over the years, the transit circle will be able to measure the motions of the stars, and so also improve our knowledge of the rotation of our Galaxy.

The two Swedish telescopes together cost about £1 million, and the Anglo–Danish transit circle cost a similar sum. As well as its half share in the latter, Britain has a much larger stake in the observatory – three reflecting telescopes to the tune of £22 million. The largest is the 4.2-metre William Herschel Telescope. Its mirror and metalwork have been completed, at the Newcastle works of Grubb

The Anglo–Danish transit circle is a small refractor swinging north–south between two piers. It can measure the positions of 1000 stars every night

Parsons, and workmen on La Palma have already poured the concrete for the walls of its dome. It should start observing in 1987.

The giant telescope will have two smaller companions, to ease its workload by studying objects which do not require its mighty eye. In the winter of 1983/84, these two British telescopes should be able to "see" the stars and other objects, recording the images on a TV camera at the focus of each telescope. After this moment of "first light", astronomers and engineers from the Royal Greenwich Observatory must check out the various light detectors, spectrographs, automatic guiding mechanisms, and their associated electronics and computers. The telescopes should be ready for regular observing in the spring of 1984.

The Isaac Newton Telescope is the larger of these telescopes, and the largest instrument so far on La Palma. Originally erected at the site of the Royal Greenwich Observatory in Sussex in 1967, the telescope and its mounting have been altered and improved to do

The building housing the Isaac Newton Telescope is also the headquarters for British astronomers on La Palma

justice to the excellent site, and to incorporate the latest developments in telescope control and instrumentation. It has a better mirror. The original 98-inch mirror was made of glass similar to Pyrex. The shape of even such a low-expansion glass, however, alters slightly as the air temperature changes, and this deforms the focused images of the stars. The new mirror is made of Zerodur, a glass-ceramic material that does not alter in size when heated over normal temperature ranges. It is also slightly larger, turning the Isaac Newton Telescope into a 100-inch telescope. The telescope's equatorial-type mounting has its axis pointing to the north pole of the sky (near the Pole Star), and moving the instrument from the latitude of Sussex to that of the Canary Islands has meant tipping the mounting up by a further 23 degrees.

The dome housing the Isaac Newton Telescope is atop a large windowless building which forms the headquarters for British astronomers at the observatory. It contains a vacuum tank for re-coating the mirror with its reflecting film of aluminium; mechanical and electronic workshops, and darkrooms; and a library and rest rooms for the astronomers. The building will remain the British headquarters even after the William Herschel Telescope is complete; the building for the major telescope will in fact be a lot smaller, as it will not duplicate many of these facilities. And the Isaac Newton Telescope itself will certainly not become redundant when its larger brother comes into operation, for astronomers will not always require the giant eye of the William Herschel Telescope, and can undertake many observations – especially of spectra – with the Isaac Newton Telescope.

The third British telescope has a mirror only 1 metre in diameter. Optical specialists Charles Harmer and Charles Wynne have developed a new mirror system for the telescope, which enables it to take unusually wide-angle photographs of the sky, up to 1½ degrees across (three Moon-breadths), with no distortion at the edges of the field. In a conventional telescope, the large primary mirror which collects all the starlight is shaped like part of a parabola. It collects and focuses light onto a secondary mirror which forms the image that astronomers (or their photographic plates) see. This mirror normally is curved like a hyperbola. In contrast, the 1-metre telescope's secondary mirror is spherical in its curvature, and it directs the light through a special set of correcting lenses. It is this combination that allows astronomers to take such distortion-free photographs.

Often, objects that are very bright sources of radiation at radio or

X-ray wavelengths are very faint sources of visible light. By examining the photographs taken using the 1-metre telescope, astronomers can locate such faint objects very accurately, because each plate is large enough to contain several bright stars whose positions are known precisely from the work of the Anglo–Danish telescope next door on La Palma, or from the position-measuring satellite Hipparcos. The faint, distant galaxies and quasars that the telescope records will also provide a fixed reference frame, against which astronomers can measure how the stars in our Galaxy (usually taken as a "fixed frame") are really moving.

A telescope mirror collects and focuses the light from the sky, like the lens in our eyes, but a telescope also needs a "retina" to record the light. Traditionally, astronomers have used photographic plates. But photographic emulsions are not very sensitive: they record only a few per cent of the light falling on them. Now astronomers have a far more sensitive retina: the charge-coupled device (CCD). These are light-sensitive silicon chips which will respond to two photons of light in three. So instead of spending a couple of hours in exposing a photographic plate, astronomers can get the same information in just a few minutes – or, with a longer exposure, they can record objects too faint to appear at all on a photograph.

Several companies make CCDs for use in television cameras, and astronomers now use such CCDs regularly. But these chips contain only as many lines as a TV picture, and astronomers would like to record fine details over a wider field of view. Hence GEC in the United Kingdom is now working with astronomers to produce a high-quality CCD, with 1500 lines.

Most of a large telescope's time is spent, not in taking direct pictures of the sky, but in breaking up the light into a spectrum. The Isaac Newton Telescope has a variety of spectrographs, some equipped with CCDs to record the spectrum. Astronomers using the telescope also have a powerful detector called the Image Photon Counting System at their disposal. Invented 10 years ago by Alec Boksenberg, now director of the Royal Greenwich Observatory, the device brightens up the spectrum with a commercially available image intensifier, whose output is viewed by a TV camera. Normally, the intensified view on the image-tube screen is marred by electronic noise, which looks like "snow" on a TV screen. The Image Photon Counting System's great strength is a computer system which analyses each spot on the screen and can tell whether it is the image of a photon of light, or simply a bit of noise. It throws out the latter, and builds up in its memory a noise-free spectrum.

Thorn-EMI is making the really high-quality image tubes needed for the far-seeing William Herschel Telescope.

Although it is not the biggest, when it is equipped with such light detectors the William Herschel Telescope will be the world's most powerful telescope. The Soviet 6-metre telescope has old-fashioned light detectors, and a thick glass mirror that distorts severely with temperature changes. The American 5-metre telescope is still a leading instrument, although it is over 30 years old – but its night sky has doubled in brightness since it was built as the lights of neighbouring towns have spread up towards Palomar Mountain.

The British telescopes are the first to be designed from scratch to be operated by remote control. Once the telescopes are in operation, an astronomer need not travel to La Palma to use them. He – or she – can sit in a control room in medieval Herstmonceux Castle (the headquarters of the Royal Greenwich Observatory) in Sussex, at a console which duplicates the set-up in the control room in the telescope dome, and operate the telescopes with signals sent along ordinary telephone lines. One television screen will display the telescope's view, while another screen shows a simple version of the data (a spectrum, for example) that it is recording. The magnetic tape containing all the astronomer's data can be sent back by air, to arrive as quickly as he could have brought it back himself. Astronomers at Herstmonceux tested remote operation in 1982, when they successfully controlled an American telescope in Arizona. As well as saving on travel costs, the system should make the observing programmes more flexible.

The international character of the Roque de los Muchachos observatory has expanded since the agreement between the original four countries, with two other countries taking shares in the British telescopes. The Netherlands is supplying 20 per cent of the cost and research effort on the three telescopes, in return for the same proportion of the observing time, while the Republic of Ireland is paying for the use of the 1-metre telescope on 27 nights per year. A consortium of Nordic countries – Sweden, Denmark, Norway and Finland – is considering a 2.5-metre reflector, while the Italians are testing it as the site for a 3.5-metre telescope.

But astronomers will always want to catch more of the faint light from the heavens, to study the farthest objects of any particular kind: planets of other stars; individual stars in distant galaxies; galaxies and quasars on the "edge of the Universe", where we look back in time almost to the moment of creation in the Big Bang. British astronomers have proposed two further telescopes to bring

The steelwork of the huge William Herschel Telescope is tested in the works at Newcastle-upon-Tyne. It stands 18 metres high and weighs 190 tonnes. The vertical fork "altazimuth" mounting makes the engineering simpler, and the more complex motions required to follow the apparent motion of a star are under computer control

these objects into view. The Royal Observatory at Edinburgh operates a wide-angle telescope in Australia, of the Schmidt design (with a thin lens at the top, as well as a large mirror at the bottom), and has suggested constructing a larger one – as big as the Isaac Newton Telescope – to scout out faint but interesting objects. Astronomers could then investigate these with a multiple-mirror telescope designed by the Royal Greenwich Observatory. Its six huge mirrors, each larger than any telescope existing today, would collect as much light as a single mirror 18 metres across. It could see objects 20 times fainter than the limit of the William Herschel Telescope.

In any event, the William Herschel Telescope, and its international supporting cast of telescopes, will ensure that La Palma remains in the front rank of observatories.

37

The first Multiple-Mirror Telescope

KEITH HINDLEY

13 April 1979

The science fiction idea of a telescope with six separate mirrors has become reality in Arizona. Its six mirrors together make the Multiple-Mirror Telescope on Mount Hopkins effectively the world's third largest telescope – at a fraction of the cost of a similar telescope with just one huge mirror.

The Multiple-Mirror Telescope (MMT) is the prototype of a new generation of large astronomical instruments of radically new design. With a light-gathering power of a 4.5-metre telescope, this new instrument utilises six 1.8-metre mirrors on an altazimuth mounting housed in a small lightweight revolving building. Despite the costly development of engineering, electronic and optical design innovations, the telescope has been built for less than a third of the likely cost of a conventional 4.5-metre single-mirror telescope and dome.

The MMT is a joint project between the University of Arizona and the Harvard–Smithsonian Center for Astrophysics, and the design has evolved in several stages. The six identical telescopes are mounted symmetrically about a central axis which allows their images to be brought to a common central focus by using three additional reflections in each telescope. This design avoids the weight, long construction delay and cost of a massive single mirror. The precise alignment of the six optical beams is maintained by automatic sensing and control in a lightweight support structure. The six-mirror assembly is carried on a simple altazimuth mounting. In this form of mounting, the telescope system is pivoted about a horizontal axis which is carried on a large fork rotating around its vertical axis. It avoids the immensely massive fork, yoke or horseshoe which would be needed in a traditional equatorial (tipped-up)

The Multiple-Mirror Telescope on Mount Hopkins, Arizona, is effectively the third largest optical telescope. Its six 1.8 metre mirrors, whose common support structure shows within the rotating building, are equivalent to a single mirror measuring 4.5 metres across

mounting large enough to accommodate the 6.4-metre-wide optical system. The problems in tracking astronomical objects using an altazimuth have now largely been solved in the recent design of large radio telescope mountings, and so the system was adopted in spite of the optical astronomers' inborn prejudice against it.

Since the focal length of the telescope is that of a single 1.8-metre instrument, it could easily fit inside a conventional hemispherical dome only 18 metres across. However, such a small dome would be seriously weakened by the 7-metre wide slit required to accommodate the width of the optical system. A square-shouldered building was a better proposition. It soon became clear that by making the top half of the structure move round with the telescope, the unused volume could be filled with a control room and laboratories without obstructing the telescope. Having progressed thus far, the logical next step was to make the whole building, weighing some 700 tonnes fully fitted out, move round with the telescope. The result is a remarkably small, extremely strong structure resembling a barn rather than a conventional astronomical observatory. The new telescope is embedded within a working laboratory with research facilities within a few metres at all levels, while ancillary equipment, such as spectrographs weighing up to several tonnes, can be added and removed with ease from the telescope. This latter benefit should prove of particular value as the complexity and weight of detecting equipment used in astrophysics increases.

The Multiple-Mirror Telescope bristles with design innovations. The heart of the instrument consists of the 1.8-metre primary mirrors of a lightweight hollow design weighing only 550 kilogrammes each. Front and rear silica plates, each 2.5 centimetres thick, are braced by an array of internal silica stiffeners in an egg-crate pattern of squares sealed with an outer wall. These components were fused together to form the blank from which each mirror was made. After inspection the mirrors were heated up for 70 hours and then raised rapidly to 1550° C for 20 minutes. This softened the silica and allowed the blanks to slump into moulds which gave their front surfaces approximately the correct curvature. After annealing to remove any stains, the technique produced lightweight primary mirror blanks which required the minimum of glass removal during the grinding and figuring stages. Tests have shown no surface indication of the egg-crate interior and the blanks have performed as well as normal solid silica mirrors.

One of the unique features of the MMT construction has been the batch production of large quantities of optics. In all, just under 100 optical components are required by the telescope and its guidance system. Seven primary mirrors were fabricated to provide a spare which would be available if one of the six mirrors in use is damaged or needs resurfacing. The primary mirrors have to be exact duplicates to within stringent tolerances. The combination of the six optical beams to produce stellar images less than 0.7 arcseconds across requires six mirrors with focal lengths identical to within 3 millimetres, and image scales identical within 1 in 1000. This has been achieved by taking the seven main mirrors in turn through standardised steps of grinding and polishing.

The MMT has two sets of secondary mirrors, designed for use in visible and infrared wavelengths. Although both sets have identical shapes, the infrared set is 2 centimetres smaller in order to reduce the amount of stray infrared radiation from the telescope and its mounting, one of the main limiting factors in the efficiency of infrared work. The production of 12 identical hyperbolic surfaces for these secondary mirrors would have taxed the resources of the most competent optical workshop. However, much of the work on these mirrors has been performed by a grinding machine controlled by a computer program specially developed for the project. The machine produces secondaries which require only hand-finishing, using two extremely accurate standard test mirrors.

The telescope can be operated in two different modes. In the first, the six light beams are brought together near the centre line by plane mirrors where a six-sided beam combiner produces a single image at what is the equivalent of the Cassegrain focus of a normal telescope. Here star images are expected to be around 0.7 arcsecond across under perfect conditions with an undistorted field of view covering 5 arcminutes. The telescope can clearly never be used as a wide-angle patrol instrument, but this field compares favourably with that available from, for instance, the Palomar 5-metre telescope.

In the second mode, the secondary mirrors are moved 10 millimetres farther from the primaries and the light beams are brought together at the central axis. This is an ideal arrangement for using multiple cooled detectors or for a particularly efficient spectrograph specially designed for the MMT.

The support of the main mirror has always presented problems in the design of large telescopes. In the MMT, however, the problems are only those of mounting 1.8-metre mirrors, minimised by the altazimuth mounting which tilts the mirrors only in one plane. The rear of each mirror is supported on a neoprene air bag while the edge is held by two roller chains located on the front and back mirror plates. As the telescope tracks in altitude, an attitude sensor meters the correct amount of air to the bags to ensure proper support, while free-hanging weights act through levers to provide the roller chains with the correct tension. A mirror and its encasing cell weighs about 2 tonnes and can readily be removed from the optical support structure and taken down the mountain for repair or resurfacing. Cleaning is done *in situ*.

The star-guidance system for the MMT is a 75-centimetre short-focus reflecting telescope with hyperbolic primary and secondary mirrors (Ritchey–Chrétien optics). This gives a field of view of about 1 degree across, and the telescope has a guidance detector which can utilise a guide star anywhere within this field. The whole MMT can be guided smoothly to tolerances well within the expected 0.7 arcsecond resolving power so the telescope motion does not blur the images. It uses a specially written computer program which generates the variable drive speeds in both axes required by the altazimuth mounting. Guidance is possible anywhere in the sky except for a small region overhead which would require excessive tracking speeds.

The MMT faces the same problems as all observatories in controlling the "seeing conditions" close to the telescope. Temperature gradients in the air can destroy image quality, and special efforts have been made to eliminate them in the MMT which has so much in the way of heated rooms and motors in close proximity to the telescope. Heat sources underneath the telescope have been eliminated by extending the observing floor up to the fork arms of the mounting and sealing the gap with flexible padding. This floor contains cooling coils and is kept at a temperature just below the expected night-time temperature. The walls of the laboratories are clad in a foam-sandwich panel with metal skins which rapidly assume the temperature of the surrounding air. The telescope itself poses the greatest problems through the drive motors and electronics and the slowness of the metal optical support system to cool at night. The frame again employs active cooling to reduce its temperature while a novel air-conditioning system greatly reduces the effect of the drive motors. Air is drawn from the observing area down through the yoke and pier into the basement chamber. The air enters through openings in the top of the yoke arms, carrying away heat from the altitude motors on each side, while exhaust gases from the laboratories are also dumped into the basement. Special ducts transport this waste hot air 60 metres down the northeast side of the mountain before releasing it. The rotating building itself is sealed at ground level by a liquid trough which not only ensures the efficiency of the waste air system but also keeps out dust, dirt, insects, snakes, scorpions and rats, with which Mount Hopkins abounds.

Ideally, the astronomers using the MMT will have remarkably little contact with it. They will sit in a comfortable control room acquiring astronomical objects and supervising guidance with a television system while monitoring the output of their recording equipment on a computer display. An observer's cage is provided at the equivalent of the Cassegrain focus and equipment can be supervised there if need be. Two platforms on the fork mounting (at the Nasmyth foci) can be fed with the combined light beam if particularly cumbersome or heavy detecting equipment is in use, and each platform can be fitted out with an insulated box. Even more bulky equipment can be operated from the rooms adjacent to these platforms since the wall facing the telescope can be removed. The mounting is extremely strong and rigid and can carry very heavy instrument loads. Within metres of the instrument are wet and dry darkrooms and a cryogenics laboratory for handling materials such as liquid helium, popular for cooling infrared detection equipment.

It is clear that the MMT is far more than just another large telescope. It will be a test bed for investigating the soundness of the MMT concept and will furnish operational experience which will be invaluable when a decision is taken whether to proceed with a larger optical-infrared instrument in the 10–15-metre class. If the concept is vindicated then the multiple-mirror system may become the basis for most future large astronomical telescopes.

38

The shape of telescopes to come

NIGEL HENBEST
26 March 1981

Optical astronomers are investigating new ways of building bigger telescopes, to reveal fainter and more distant stars, galaxies and quasars. Two American universities are already constructing telescopes larger than any existing today: the Texan telescope has a very thin mirror 7.6 metres across, while the Californian instrument is a 10-metre dish made of mirror segments fitted together. But future telescopes may be even more revolutionary.

The first astronomical telescope, in 1609, was Galileo's "optick tube" with a lens only 4 centimetres across. Ever since astronomers have been greedy for bigger ones. *Refracting* telescopes – using lenses – grew steadily from Galileo's original version to the huge Yerkes refractor, opened in 1897, which, with its 1.01-metre (40-inch) lens is still the largest in the world. *Reflecting* telescopes – using mirrors – have grown from Sir Isaac Newton's 2.5-centimetre toy to the giant 5.08-metre (200-inch) mirror on Palomar Mountain and the Soviet 6-metre (236-inch) instrument in the Caucasus. After a lull of 30 years, astronomers are not beginning to design even larger instruments – their Holy Grail being a 25-metre reflector: the Thousand-Inch Telescope.

What astronomers have been crying out for throughout this century is "more light". Double the size of a mirror or lens, and the collecting area increases fourfold. As most astronomical objects are faint, more light means more information, more quickly. Alternatively, one can "see" similar types of object twice as far off in space.

Refracting telescopes reached a limit with the Yerkes instrument, whose 1-metre lens weighs almost a quarter of a tonne. A larger lens would sag under its own weight, and the resulting distortion would spoil the images the lens produces. Light travels through a lens, and so one must use a large piece of glass free from all imperfections; and the sheer thickness of a lens means it absorbs an uncomfortable fraction of the light passing through.

So, since the turn of the century, "large telescope" has meant a reflector. The front face of a curved mirror focuses light as accurately as a lens can and the light does not pass through the mirror, so its thickness and internal imperfections do not degrade the image. Most important, a mirror can be supported uniformly from behind to prevent it from sagging. In principle, there is no limit to the size of reflector that can be built.

The driving spirit behind the modern large reflectors was the American astronomer George Ellery Hale. While still in his twenties, Hale had persuaded the Chicago tycoon Charles Yerkes to fund the giant refractor which now bears his name. Moving to the Mount Wilson Observatory near Los Angeles, Hale built the first 1.5-metre (60-inch) reflector of modern design and followed it by the famous 2.5-metre (100-inch) Hooker reflector – again funded by a wealthy businessman, John D. Hooker. To probe deeper into space, Hale acquired funds from the Rockefeller Foundation for an unprecedented technological leap to double the size of telescopes in one bound, by constructing a 5-metre (200-inch) instrument on Palomar Mountain.

The 5-metre telescope was completed after Hale's death, and is fittingly known as the Hale Telescope. Since its inauguration in 1948, it has reigned supreme. Many modern telescopes have been constructed since then, but all (bar one) are smaller. Typical of the new generation of telescope, in the 3.5–4-metre size bracket, are the 3.9-metre Anglo-Australian Telescope in New South Wales, the 4-metre telescopes at Kitt Peak, Arizona, and Cerro Tololo in Chile, the 3.6-metre European Southern Observatory Telescope at La Silla, also in Chile, and the 3.6-metre Canada–France–Hawaii Telescope in Hawaii.

The only rival in size to the Hale Telescope has been the Soviet 6-metre instrument at the Zelenchukskaya Astrophysical Observatory in the Caucasus. Opened in 1976, after years of delay, the telescope is generally reckoned to be no better than the American 5-metre reflector. The mirror produces poor-quality images, and the Soviets lack essential sophisticated back-up techniques and detectors.

Increases in the size of telescopes have been limited by enormous strides in electronic techniques for detecting light. Even as the Hale Telescope reached completion, the electronics industry was perfecting the first generation of light-sensitive detectors that would increase the power of existing telescopes at a fraction of the cost of building a larger instrument. The Hale Telescope today, with a modern electronic CCD detector, is equivalent to a 50-metre telescope equipped with photographic plates of 1948 vintage. Astronomers have effectively increased the size of telescopes tenfold by using the new detectors.

The sensitivity of detectors can go little further, though, now that they can record 70 per cent of the incident photons. In the past few years, therefore, astronomers have begun to look again at the possibility of constructing a new generation of telescopes with larger collecting areas, to catch more photons from the sky.

The main problem with large reflectors is simply one of cost. Aden Meinel, of the University of Arizona, has investigated the cost of major American telescopes in the size range 0.7–5 metres, and finds that the cost increases very steeply with size. The total cost, including the telescope mounting, the protective dome, and site development, is not directly proportional to the size of the telescope, but increases exponentially: the cost is proportional to the diameter of the mirror raised to the power of 2.6. Extrapolating, one finds that a 25-metre telescope would cost £1000 million. For the same sum, one could build 70 5-metre telescopes – and this would be much more effective for astronomy, as the collecting area of a 25-metre reflector is only 25 times greater than that of a 5-metre mirror.

Much of the cost, however, goes not into the telescope itself, but into its mounting and dome, and recent years have seen ways to cheapen both considerably. Any telescope must follow stars across the sky, as they appear to turn about the pole of the sky as a result of the Earth's rotation. The traditional equatorial mounting has one axis pointing at the pole of the sky. During the night the telescope is simply turned at a constant rate about this axis. It is very convenient mechanically, for it employs a simple, constant-speed motor which can keep the telescope tracking with a pointing accuracy of 1 arcsecond or better. But the equatorial mounting is an engineering nightmare. It involves swinging hundreds of tonnes of asymmetrically shaped metal about an inclined axis. The constantly changing stresses mean the mounting must be exceptionally strong – and hence exceptionally expensive.

The possibility of computer control has now altered this view of mountings. If a telescope is mounted on vertical and horizontal axes (altazimuth mounting), the engineering is much simpler and the mounting consequently cheaper. It then needs to be driven about both axes simultaneously, and at different and varying rates, depending on the star's position in the sky. A computer, however, has no difficulty in calculating the necessary drive rates continuously, and in controlling the drive motors to any degree of accuracy. The Soviet 6-metre reflector was the first large modern telescope built on an altazimuth mounting, and one of its undeniable achievements has been to prove that such mountings do work.

The other major cost of a telescope is the protecting dome. The cost of the dome increases as the cube of the telescope's length – and so, for a fixed design, as the cube of the mirror's diameter. The dome for a 4-metre telescope costs as much as the telescope within; for larger sizes the dome is more expensive than the telescope!

One simple answer is to shorten the focal length of the telescope by using a mirror of greater curvature, thus making the tube shorter and the dome smaller. Modern telescopes are already following this trend: the Hale 5-metre instrument has a focal length of 16 metres, while the 4-metre Mayall telescope at Kitt Peak focuses starlight in only 10 metres.

The alternative to shorter focal lengths is once again to dispense with tradition. A conventional dome is big enough for the telescope to swing freely around inside, because it is difficult to relate the rotation of the dome around its vertical axis with the motion of the telescope about the inclined axis of the equatorial mounting. With an altazimuth mounting this problem disappears. The "dome" need move only about the same vertical axis, at the same rate, as the telescope itself. Hence the building need be only marginally larger than the telescope, and it can be square or oblong instead of circular, thus easing problems in construction. The University of Arizona and the Harvard–Smithsonian Center for Astrophysics have pioneered such a building for their Multiple-Mirror Telescope on Mount Hopkins.

The problems in cost thus boil down to the telescope itself, and ultimately depend on the weight of the main mirror. A large mirror for a telescope can be made lighter by making it thinner, but it is then more flexible and can deform badly. In addition, it is impossible to grind and polish a thin mirror more than 7 or 8 metres across. A 25-metre telescope needs a fresh approach (Figure 1).

A dozen design teams have investigated a variety of possibilities, questioning conventional ideas at every stage. They presented their proposals at a conference in Tucson, Arizona, in January 1980. And a weird bunch of designs they were. Some of the most bizarre-

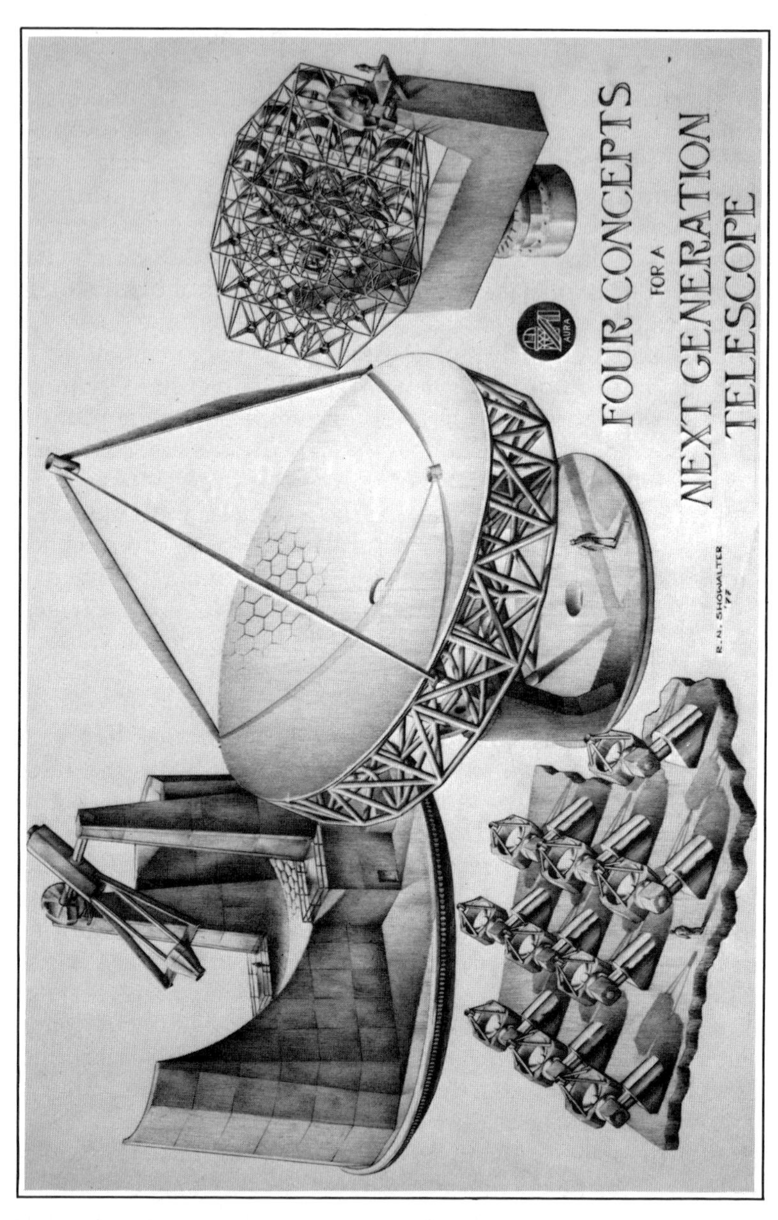

Figure 1 *Four designs for a telescope of the not-too-distant future, as envisaged by American astronomers in the late 1970s. The "singles array" (lower left) and the detector-scanned turntable design (top left) have not found as much favour as the two rather more conventional concepts. In the centre is a huge reflecting dish made of many small segments; on the right is a scaled-up version of the Multiple-Mirror Telescope.*

looking have now been dropped – unfortunately in some ways, because one always has a sneaking hope that something completely unorthodox may prove better than a conventional design. Among the weird designs were a fixed concrete bowl lined with mirrors, and a mirror-lined "shoe"; in both of these designs, the detector at the focus moves to counteract the Earth's rotation. There were a couple of two-mirror designs, where a large flat mirror is tilted to reflect light to a fixed curved mirror – one design placed the mirrors in a horizontal box, earning it the title of the "rotating mailbox".

With a general mood of optimism about a New Technology Telescope emanating from Tucson, the proposals were put to the Field Committee of the National Academy of Sciences. Under astrophysicist George Field, the committee was charged with recommending astronomy policy in the USA for the next decade. Its guideline has been the cost of the Very Large Array radio telescope recently completed in New Mexico (about £60 million). This is not sufficient for a 25-metre optical telescope, but it will cover a 15-metre instrument – which would be very useful in its own right, and a test bed for the new technology needed for a 25-metre telescope as a longer-term goal.

The Field Committee changed the emphasis somewhat from that prevalent at the Tucson conference. The committee's most active lobbyists have not been the optical astronomers but their infrared colleagues. The New Technology Telescope would be as useful for "near" (short-wavelength) infrared observations as for optical work, and in some ways better: atmospheric "twinkling" decreases at longer wavelengths, and in the near-infrared the size of the blurred images is down to ½ arcsecond. It was largely infrared considerations that ruled out the bowl, shoe and mailbox designs. In the two fixed-mirror designs, the moving detector would pick up reflections from different parts of the mirror which may differ in their infrared reflectances; the mailbox involves two reflections, and every additional mirror adds noise to the infrared signal.

One of the three front-runners to emerge from the Tucson conference has also been eliminated. This was the "singles array". To make the equivalent collecting area of a 25-metre dish, the concept employs a hundred individual 2.5-metre telescopes, each in its own dome, with all the outputs combined at a central laboratory. It has a number of advantages: each telescope requires no new and unproved technology, and their costs would fall by mass production; the telescope would start producing results as soon as the first few units were installed, which could be within only a few years; and it is financially flexible, because one can add telescopes as money becomes available.

Infrared astronomers have objected strongly to this proposal, however. It would be too expensive to equip each individual telescope with a detector, and the alternative is to send the radiation directly to the central laboratory by a series of reflections. But the number of mirrors involved would rule out infrared work, and the precision required to overlap images from separate moving telescopes to a precision of 1 arcsecond is probably beyond the limits of present technology – the singles array is not the "low-technology" device its proposers have suggested.

The Field Committee is recommending a comparison of two other types of new telescope, both springing from design studies that date back many years. One of these is a development of a "new technology" telescope that is already in use. Completed in 1979, the Mount Hopkins Multiple-Mirror Telescope (MMT) represents the greatest break with astronomical tradition this century. It has an altazimuth mounting, and a square rotating protective building; but the telescope itself is even more revolutionary. It consists of six separate 1.8-metre mirrors held in a common framework. The light from each is combined by mirrors at a common central focus.

In terms of total collecting area, the MMT is the third largest telescope in the world, equivalent to a single mirror 4.5 metres across. Its designer, Roger Angel, sees little problem in scaling it up to 14-metre equivalent aperture for the national project envisaged by the Field Committee. He and Nick Woolf are submitting detailed designs for the required eight-mirror telescope, each mirror being 5 metres in diameter. They intend to keep the mirrors light in weight by using a honeycomb construction: a thin curved front plate (polished to the correct mirror surface on top) is bonded to a thin back plate by narrow spacers, as used for the existing MMT mirrors.

Groups at the University of California and Kitt Peak National Observatory have been independently examining the alternative, a single mirror 15 metres in diameter. Because it is impossible to make and polish so large a piece of glass, the designs call for a central continuous disc about half that size, surrounded by dozens of smaller mirrors fitted together. These small segments are supported individually from behind, and can be tweaked to keep the mirror surface correctly shaped as the telescope tilts.

But there are two major problems in producing these mirror segments. They are parts of a paraboloidal or hyperboloidal surface, well away from the central axis – so each segment in itself must have a strange asymmetrically curved surface, which is impossible to figure and polish by conventional techniques. Jerry Nelson of the University of California has solved this particular problem very neatly. He bends each segment in a carefully calculated way, then grinds and polishes it to a shallow spherical figure. When the mirror is released, it springs back to the desired asymmetrical shape.

Infrared astronomers require that the surface of such segmented mirrors be continuous, with no gaps between the segments. The mirror segments thus cannot be circular in outline, but must be either hexagonal or slightly rounded rectangles, to fit neatly together. The mirror-makers are pretty sure that they cannot polish mirrors with corners, so the two American teams intend to cut circular mirrors to shape.

Shortage of funds may delay the completion of the New Technology Telescope, especially as the Field Committee placed it third in order of priority, after the orbiting Advanced X-ray Astrophysics Facility and an array of radio telescopes across the country. Even if there is no national project, the United States will almost certainly have one or more telescopes larger than the Hale 5-metre reflector by the end of the century. The University of California had sunk almost £1 million into design studies for a segmented-mirror telescope with an aperture of 10 metres even before the larger national project was proposed. Jerry Nelson's group there has the full backing of the university, which may proceed with its own 10-metre instrument anyway: the estimated cost is around £30 million.

Even more probable is a 7.6-metre telescope proposed by the McDonald Observatory. This would be a conventional telescope, which takes the construction of a single-piece ("monolithic") mirror to the limit – in the Texan mode of simply making it bigger than anything existing. The mirror would be kept lightweight by its extreme thinness. The pioneer in thin monolithic mirrors is the UK's Infrared Telescope on Hawaii, which has been astoundingly successful in preserving an accurate figure. It has a sophisticated support system which corrects for any distortions in the mirror, which is 3.8 metres wide, and only 24 centimetres thick (on average). The Texans plan a mirror whose thickness is only 2 per cent of its diameter!

The European Southern Observatory has also been investigating designs for a 16-metre telescope, and the Russians are discussing a 25-metre segmented-mirror telescope. The next 20–30 years promise to be exciting for the builders of telescopes, and for everyone who loves watching technology being stretched to its limits.

39

The honeycomb route to huge mirrors

"TECHNOLOGY"

10 November 1983

American telescope-makers have built a prototype hollow mirror which is cheap and only one-fifth as heavy as a conventional mirror. It is the first step in a plan to make the world's largest telescope mirror, 8 metres in diameter.

All big optical telescopes now are reflectors, collecting light with a large curved mirror. The surface is ground in the front of a glass disc, the mirror blank, which is then polished and coated with a reflecting layer of aluminium.

There are two problems with simply scaling up conventional mirror blanks. The mirror's weight increases as the cube of its diameter, so that doubling the size of a telescope produces a mirror eight times heavier – and the entire telescope and its mounting must be made far stronger (and more expensive) to support the extra weight.

The second problem is that the mirror expands and contracts with temperature changes from day to night. These small fluctuations in size can seriously affect the accuracy of the mirror's surface. Astronomers must wait for the telescope to cool at night. To circumvent this problem, most recent telescope mirrors have been made of glass-ceramics which have a negligible coefficient of expansion. But they are much more expensive than "ovenware" glasses such as borosilicates and aluminosilicates.

Roger Angel and his team at the Steward Observatory, Arizona, have pioneered a different approach. They reduce the mirror's weight by making it like an empty honeycomb. Between the front plate (to be polished and coated) and the back face, the mirror is largely hollow, and consists merely of the walls surrounding hexagonal-shaped holes. The final mirror weighs only about one-fifth as much as a solid mirror.

The honeycomb spaces in this 1.8-metre mirror make it only one-fifth as heavy as a solid-glass mirror

The honeycomb structure has another advantage. It allows astronomers to blow air into the mirror, and so quickly bring the whole mirror down to the temperature of the night air. An 8-metre glass mirror would not acclimatise in a whole night, but a hollow mirror would cool in just an hour. As a result, Angel's team did not need to make its mirrors from expensive glass-ceramic, but can use ordinary borosilicate glass.

The team made a prototype mirror blank, 1.8 metres in diameter, by melting chunks of raw glass in a circular mould, which contained hexagonal blocks of ceramic fibre mounted on bolts made of silicon carbide. When heated in a furnace to 1200 degrees C, the glass melted and ran over, under and around the blocks. The smooth flat top became the front of the mirror blank. After cooling, the team removed the bolts at the back (thus leaving holes in the back plate which will ventilate the mirror), and applied high-pressure water jets to blast out the ceramic fibre blocks.

The mirror's front surface must now be ground to a parabolic shape. This process is time-consuming, and for the proposed 8-metre telescope it would mean grinding away 10 tonnes of glass. So the engineers are trying a novel method on a second 1.8-metre blank. They are spinning the entire oven about a vertical axis, at a rate of 5 revolutions per minute. Such spinning should cause the surface of the glass to form naturally into the required parabolic shape. The next step is to build a revolving furnace big enough to make a mirror 3.5 metres across.

40

Speckles reveal the stars' secrets

SIMON WORDEN

27 April 1978

An ingenious system of short-exposure photography – speckle interferometry – can "freeze out" the blurring caused by the Earth's atmosphere. It can resolve close double stars, and show the size of asteroids and even the nearest stars.

The angular resolution of an optical telescope – its ability to see detail – is set by the wave properties of light. In principle, the limit of angular resolution is inversely proportional to the diameter of the main light collector, which is either a mirror or a lens. Thus a larger telescope has a finer resolution capability than a small one. This angle is about 0.02 arcseconds for large telescopes such as the Hale Observatory's 5-metre instrument in California.

Astronomical angular resolution is badly degraded by turbulence in the Earth's atmosphere. Light from distant stars arrives outside the Earth's atmosphere in a form known as a plane wave. The wave's *amplitude* tells how strong the wave is, and its *phase* denotes the location of peaks and troughs within the wave. As long as the phase for the wave is the same across the entire telescope aperture, the size of the star image is determined solely by the diameter of the telescope. However, turbulence in the atmosphere breaks up this plane wave so that the phase is constant only over about 10 centimetres distance. Thus, no matter how large the telescope, the resolution obtained is no better than that of a 10-centimetre telescope – namely about 1 arcsecond. Unfortunately, there are many problems of astronomical interest which require much better resolution than that. For example, the largest stars other than the Sun have angular diameters less than 0.1 arcsecond.

In 1970 a French astronomer, A. Labeyrie, proposed a method to circumvent this problem. He pointed out that very short snapshots

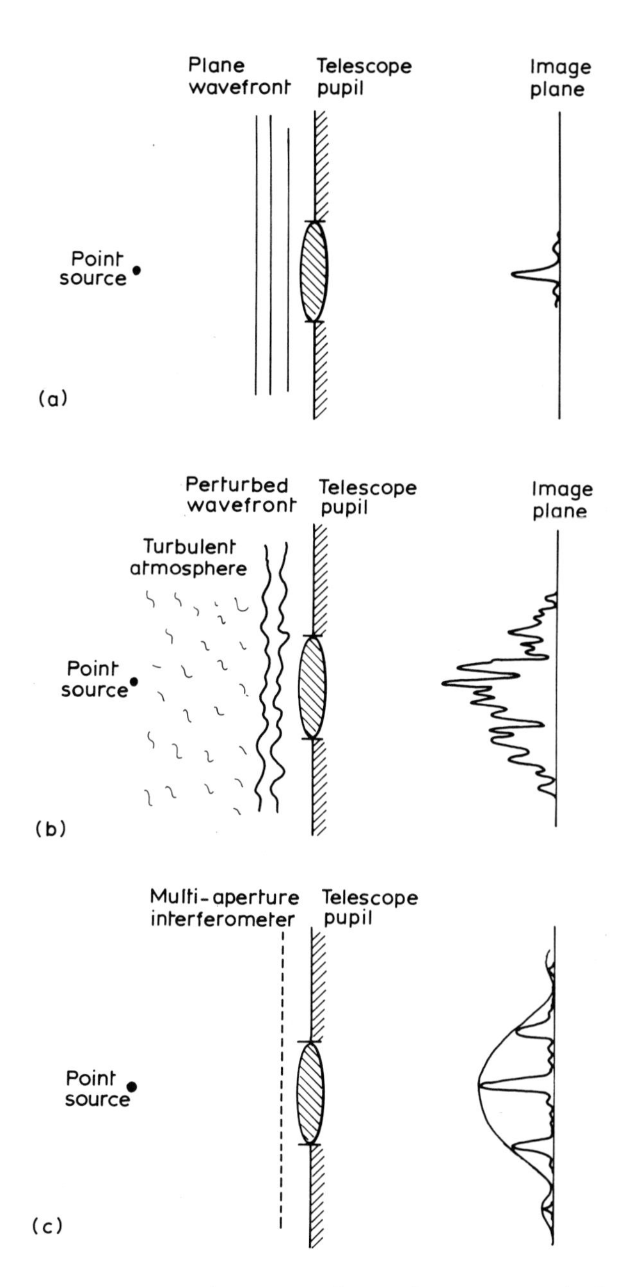

Figure 1 *Formation of an image by a telescope, in ideal circumstances (a), and with a turbulent atmosphere in front (b). The latter causes "speckles", similar to those produced by an interferometer formed by placing many small apertures over the telescope's lens or mirror (c)*

obtained on a large telescope "freeze" the rapidly changing turbulence in the Earth's atmosphere and allow significantly higher resolution than 1 arcsecond to be realised. The reason follows from optical interference theory and it is illustrated in Figure 1. Figure 1(a) represents the image formation as it would occur without turbulence. Angular resolution in this case is determined only by the aperture size of the telescope as discussed in the previous paragraph. With atmospheric turbulence we get case 1(b). As well as the much lower resolution in this image, the turbulence changes so fast that the image is rapidly altered in size and shape. This effect is visible to the naked eye as twinkling.

Labeyrie pointed out that when the exposure is 1/50 second, or less, atmospheric turbulence is frozen, in which case light arriving at one part of the telescope aperture will still be in phase with the light of a few other points in the aperture. This is known in optics as a multiple-aperture interferometer and it is shown in Figure 1(c). A. Michelson first used interferometers in astronomy in the 1920s to measure the diameters of several stars. The composite image produced by the telescope multiple-aperture interferometer consists of a large number of images of the star superimposed on each other. Each of these images is undegraded by the atmospheric turbulence, and the result is a "fly's eye" view of the star.

To carry the analogy further, these images look "speckled". Since the photographs are multiple-aperture interference patterns, the use of such data to remove the unwelcome effects of atmospheric turbulence is called speckle interferometry, where each speckle is an image of the star. In the case of a binary star each speckle is doubled. Similarly, the speckles for a big star are much larger than the speckles formed when a small star is photographed.

Fourier mathematics and Fourier processing have proved very useful in analysing speckle data. Labeyrie and his collaborators at the University of New York, Stony Brook, were the first to develop such methods for the reduction of speckle photographs. The mathematical processing of speckle images may be accomplished using direct optical instruments, as Labeyrie has done, or by using a large digital computer. Adding results from large numbers of individual speckle exposures will give highly accurate estimates of star sizes and shapes. In the case of the largest existing telescopes, the technique improves the old 1-arcsecond atmospheric limit about fifty-fold. Diameters of several stars, and separations for very close binary star systems, have been observed using these methods.

Several research groups now make regular use of speckle interferometry for the important derivation of binary star orbits and angular separations. In addition to Labeyrie, now at the Observatoire de Paris in Meudon, France, C. Dainty with his collaborators at Imperial College, London, and H. McAlister at the Kitt Peak National Observatory in Tucson, Arizona, have done work on binary systems. Binary star observations are of fundamental astrophysical interest for a number of reasons. Information on the orbit of one binary component about the other, coupled with doppler velocity observations of each component and its brightness yields just about everything an astronomer might desire to know about the component stars. Distances to the system, stellar masses, stellar luminosities, and dimensions of the binary orbit are all obtainable. These results may then be used to study the internal structure and evolution of stars with differing masses allowing comparison with theoretical predictions.

Another important use of binary stars is to calibrate the distance scale of the Universe. All distances in astronomy are based on knowledge of distances to stars in our Galaxy. The use of speckle interferometry now makes it possible to observe binary stars with separations nearly 10 times smaller than any previously observed. That corresponds to an increase in the volume of available space and number of stars by a factor of 1000.

McAlister at Kitt Peak has proposed a fascinating new use of binary star observations incorporating the high accuracy inherent in speckle interferometry results. If the major part of a binary star orbit were to be observed, it is quite possible that small "wobbles" in the orbit of one star about the other would be detected. These wobbles could then be interpreted as an indication of additional bodies in the system. The masses of such bodies could be as small as the Earth's. It is possible that new planetary systems and Earth-sized planets will be detected in this manner!

The angular diameters of stars are also being observed using speckle interferometry. Although obtainable by other interferometric methods, speckle determinations of diameters provide a useful alternative and confirmation. Star angular sizes are coupled with brightness measurements to provide estimates of stellar temperature and absolute size. Unfortunately, speckle methods are currently limited to the largest stars, the red supergiants. However, these are highly evolved ("old") variable stars and it is interesting to compare their sizes and temperatures with theoretical predictions of these quantities. In addition to Labeyrie's work in this area, I have derived angular stellar diameters at Kitt Peak.

Speckle interferometry has proved useful for determining sizes of other astrophysically interesting objects such as asteroids. Since asteroids are thought to be primordial solar system objects, their composition and structure provide clues to the origin of the solar system. In collaboration with scientists at the University of Arizona, I have used speckle interferometry to measure the angular sizes of several asteroids, providing a confirmation of alternative means for deriving their sizes. Speckle interferometry has also been applied to the closest star in the sky. J. Harvey of the Kitt Peak National Observatory and C. Aime of the Nice Observatory in France have used the new speckle techniques to study the sizes of granules on the Sun's surface and of features within sunspots. These features are smaller than 1 arcsecond, but a correct estimate of their real size is an important input into theories of energy transport in the solar atmosphere.

41

Hipparcos ties down the moving stars

ANDREW MURRAY

1 July 1982

A specially designed European satellite will measure the positions of stars far more precisely than is possible with ground-based telescopes. These results will reveal the exact distances to many distant stars, and indirectly to the farthest galaxies.

Astronomers have long regarded the "fixed stars" as the ideal frame of reference: a backdrop against which they can study many celestial phenomena. A knowledge of the exact locations of stars in this reference frame, and the relative changes in positions of all manner of celestial objects, is essential and forms the basis of the branch of astronomy called astrometry. The accurate measurement of stellar positions provides the basic data for calculating distances within our Galaxy and beyond, and for studying the dynamics of the Galaxy. The maximum distance that can be measured directly is infinitesimal compared with the size of the Galaxy, let alone the distances to other galaxies. Yet astrometric measurements form the foundation upon which is based the whole scale of cosmic distances, and hence of the age of the Universe.

In 1986, astrometry will leave the confines of the Earth, and escape the limitations that the atmosphere imposes. In that year the European Space Agency (ESA) will launch a satellite devoted to the precise measurement of the positions of celestial objects. The satellite is called Hipparcos; for *Hi*gh *P*recision *Par*allax *Co*llecting *S*atellite, and in honour of the Ancient Greek astronomer Hipparchus. It will provide a great step forward in the accuracy of astrometric measurements and will thereby have a significant impact on almost all fields of astronomy.

In recent years astronomers have increasingly had to compare the map of the sky at optical wavelengths with maps made in other regions of the electromagnetic spectrum, particularly at radio wavelengths. The technique of radio interferometry can pinpoint a source very accurately with respect to a global reference frame (one that covers the entire sky). Ostensibly, this is the frame used in optical astrometry. But radio sources are very faint at optical wavelengths and must be observed with large telescopes to collect sufficient light. Such telescopes look at only a small region of sky, which makes it difficult to relate the map of the small region recorded, for example, on a photographic plate to the global reference frame. The reason is that the frame is defined by a catalogue of stars, and few if any of these will appear in the small region covered by the plate. The optical position of a faint object relative to the global reference frame can now be specified to an accuracy of about 0.2 arcsecond, an angular measure comparable to about one-fifth of the radius of a star's photographic image. Astronomers have already pushed ground-based optical astrometry to the limits of accuracy. It is worth looking at some of these limitations, to see how astronomers will benefit from Hipparcos.

Because stars move relative to the Sun, their apparent positions gradually change over the years, an effect known as "proper motion". So the position of a star at a given time or "epoch" must usually be extrapolated from previous observations; inevitable errors in the measurement of proper motions mean that the accuracy with which a stellar position is known must deteriorate with time. There is also an overall limit to the accuracy of Earth-based observations set by atmospheric effects. Furthermore, it is not possible to observe the whole sky from any single observatory on Earth, so the global map must be built up from measurements made at many observatories, over many decades, and inevitably the map suffers distortions. By contrast the Hipparcos satellite will scan the whole sky, including a hundred thousand stars, and the reference frame defined by these stars will be free from atmospheric influences.

The optical system in Hipparcos (Figure 1) consists of a reflecting telescope of 30-centimetres apertures, in which two fields of view inclined at approximately 60 degrees are superimposed by reflection in a complex mirror situated on the optical axis. The angle between two stars, one in each field of view, is given by the so-called "basic angle" of the complex mirror, plus the small separation which is measured in the focal plane of the telescope.

The telescope will spin about an axis perpendicular to the two fields of view. In one complete rotation, which will take about 2½

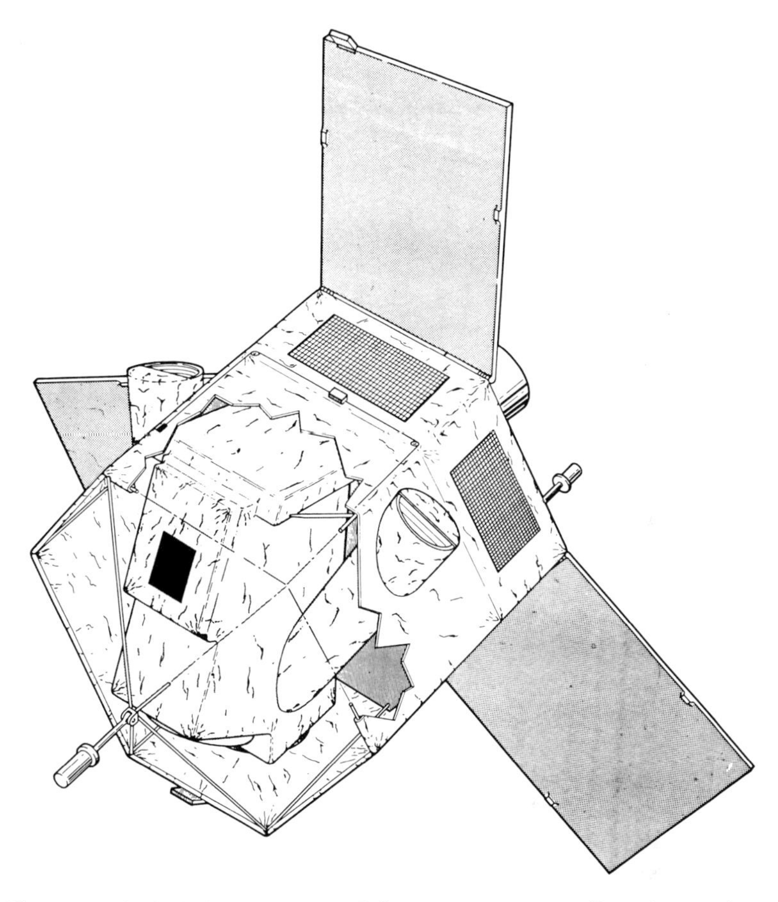

Figure 1 *Artist's impression of the Hipparcos satellite shows the two circular (half-blocked) apertures looking towards fields about 60 degrees apart. Within the shaded box, the light from both apertures falls on a double mirror which reflects the two areas of the sky onto a single focusing mirror*

hours, it will measure the relative angular distances of many stars along a great circle. A measurement is made by timing the passage of a star's image across a grid of slits in the focal plane. The spin axis will be kept at a constant inclination of approximately 40 degrees to the direction of the Sun, and will revolve around the Sun once in about seven weeks.

In this way any small region of sky will be scanned many times during the mission by great circles which intersect at well-inclined angles. Astronomers will then combine individual angular measurements to form a sky map, including the small displacements due to parallax and proper motion, using techniques similar to those used in triangulation in surveying the Earth's surface. The estimated accuracy is 0.002 arcsecond for each position coordinate, annual proper motion and parallax, in a survey consisting of a hundred thousand stars.

The stars to be observed will be selected after consultation with a large number of astronomers who stand to benefit from the improved accuracy Hipparcos will provide. In a preliminary survey some 125 European astronomers proposed more than 150 projects which require the more accurate data.

One of the most important tasks is the measurement of the stars' trigonometric *parallax*. This is the annual apparent oscillation in a star's position arising from the Earth's annual motion around the Sun. The star's distance can be easily calculated from its measured parallax angle and a knowledge of the Earth–Sun distance. The smallest parallax that can be measured from the Earth is about 0.04 arcsecond, which corresponds to a little more than 80 light years. (The light from the most distant star astronomers can measure takes a human lifespan to reach the Earth.) There are probably fewer than 2000 stars within this distance and for many of these the true distances are very uncertain. Yet these stars lie in only a small portion of the total volume of our Galaxy, and include only the very commonest types of star.

The measurement of trigonometric parallax provides the only direct method of determining distances to stars. A less direct method, however, arises from a comparison of stars' radial (line-of-sight) velocities – measured by observing the doppler shift of their spectral lines – with their apparent motion across the sky. Usually this method can only be used in a statistical way, averaging the motions of many stars. But there is a special case where this technique is very important and very precise: when the group of stars form what is known as a "moving cluster", as in the case of the Hyades cluster in the constellation Taurus.

In 1908 Lewis Boss, an American astronomer, noticed that the proper motions of many of the stars in the Hyades cluster converge towards a point in the sky some 30 degrees away from the centre of the cluster. Now, if the stars are moving along parallel lines with identical velocities, then their proper motions will appear to converge, like the apparent convergence of parallel railway tracks towards the horizon. The converse, however, is not necessarily true; if the cluster of stars is expanding or contracting, then the proper

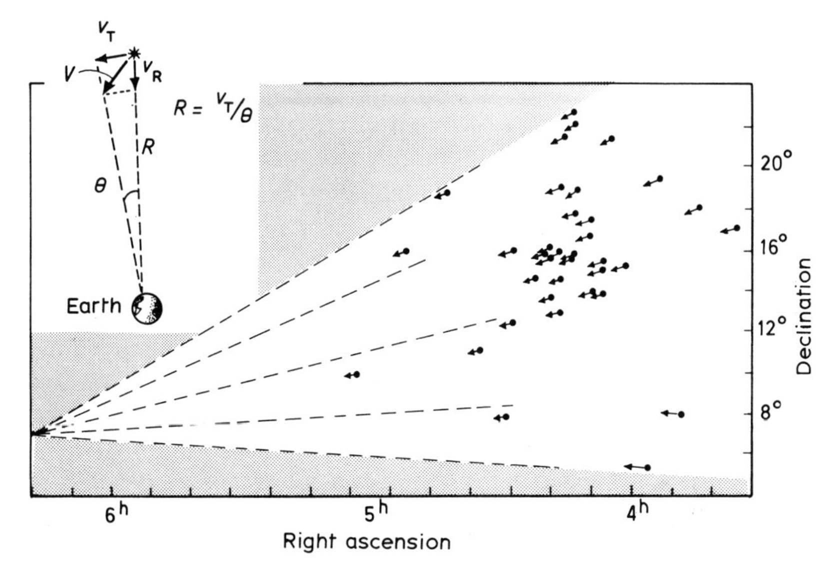

Figure 2 *The proper motions of the stars in the Hyades cluster appear to converge on a point some 30 degrees away. The length of each arrow is proportional to the star's velocity. The direction of the convergence point reveals the star's direction of motion; this and the radial velocity give the speed across the sky, and a comparison with the proper motion (in angular units) gives the star's distance (inset)*

motions will still converge. Nevertheless, if we assume that the stars in the Hyades cluster are indeed moving parallel to one another, then we can deduce from their radial velocities that the total speed towards the convergent point is about 45 kilometres per second. Knowing this total velocity, we can calculate the transverse component, perpendicular to the line of sight. The distances to the individual stars are then given by the ratio of the transverse velocity to the observed proper motion, which is also transverse to the line of sight (Figure 2).

This procedure appears to be quite straightforward but it has led to conflicting results when different selections of stars, supposedly all belonging to the cluster, are analysed. Indeed, the estimate of the mean distance to the Hyades – one of the cornerstones in deriving the scale of distances in the Universe – has varied by more than 10 per cent in recent years. The distance that is currently accepted is about 150 light years. This is as far as one can go in using ground-based observations to measure the distance to individual stars.

The observed trigonometric parallaxes of several members of the Hyades have been used to derive a mean distance to the cluster, which is in reasonable agreement with the value obtained by analysing the convergence of the proper motions of the stars in the cluster. However, Hipparcos will bring the stars of the Hyades well within the range of accurate measurements of parallax, and will reach out to several thousand more distant stars, up to more than 300 light years away.

Once astronomers have measured the distance to a star, it is possible to calculate its *intrinsic* brightness from its *apparent* brightness by assuming that the light from the star fades in proportion to the inverse square of the distance. The distance to farther stars can thus be calculated if they are of an identical type to stars in the Hyades. Most stars in fact lie on a sequence of temperature and luminosity (such that the hotter stars are the more luminous). By matching the *main sequence* of the stars in the Hyades, temperature for temperature, with the main sequence of another cluster, it is possible to calculate how much the latter's stars are dimmed by distance – and, hence, how far away the cluster lies. The method shows the Pleiades cluster – the "Seven Sisters" – lies a little more than 400 light years away, and the double cluster in Perseus at 8000 light years. The latter contains some intrinsically very luminous stars, not found in the smaller, nearby, clusters. These Cepheid variables and supergiants are among the brightest stars known in our Galaxy and are easy to identify in neighbouring galaxies. The Cepheid variables form an essential link in the cosmic distance scale from the direct measurements in the solar neighbourhood to the distant reaches of the Universe, and have a key role in estimations of the age of the Universe.

But the various links in this chain of reasoning are more tenuous than they perhaps appear to be. The standard main sequence is the combination of observations from many clusters; corrections for absorption by interstellar dust for distant clusters are uncertain, and allowance has had to be made for the change in relation between temperature and luminosity for the brightest stars of the main sequence, due to stellar evolution.

The basic problem is that the volume of space throughout which astronomers can at present make direct measurements of distance is too small to contain a representative sample of the rare but intrinsically bright stars known to exist in the general field of the sky as well as in the clusters. It is here that observations from Hipparcos will have their most dramatic impact. The fivefold increase in accuracy that Hipparcos will provide means that the volume of space over which astronomers can measure parallax directly will be increased more than a hundredfold. There will be enough red giants and stars at the bright end of the main sequence for parallax measurements to give a useful direct calibration of the intrinsic luminosity of these types of star. Furthermore, by combining new parallax data with photometric data and spectroscopic observations for large numbers of main sequence stars, it will be possible to study how the location of stars in the main sequence depends on their chemical composition.

There are many other applications for the distances and proper motions that Hipparcos will measure. For example, modern theories of the dynamics of our Galaxy suggest that spiral density waves in the galactic plane may cause deviations from the general pattern of rotation of young stars. The absolute rotations of the Galaxy can be measured directly only by observing the very small proper motions of distant stars; ground-based measurements of proper motion would not be sensitive enough to show any deviations from a general rotation. The new data will also allow astronomers to trace the galactic orbits of many more stars so as to find regions near the Sun where young stars are formed, and will provide the basic data for a much improved picture of the evolution of stars.

According to current plans, Hipparcos will be launched late in 1986 and will provide data for 2.5 years. The final results will come from a synthesis of all the data after the mission is completed, and should be available by 1991. The timescale of the project may seem rather long in comparison with other space projects, but the data from Hipparcos will supersede all the astrometric measurements that astronomers have painstakingly acquired over three centuries, and on which present ideas of the structure and dynamics of the Galaxy have largely been based.

42

Space Telescope: a dream come true

"MONITOR" and "THIS WEEK"
16 April and 2 July 1981

A large optical telescope orbiting above the atmosphere has been the astronomer's dream since space flight became reality. A joint American–European 2.4-metre telescope will make it come true in 1986. The Space Telescope will "see" stars and galaxies 50 times fainter than can be observed from the ground, and will show details 10 times finer.

Ground-based telescopes suffer from two major disadvantages caused by the Earth's atmosphere, that the Space Telescope will literally soar above. Shifting air currents in the atmosphere make the image of a star (or any astronomical object) constantly wobble about, jumping around its true position hundreds of times per second, and by angular distances of up to 1 arcsecond. During an observation, therefore, a point-like source is blurred into a disc 1–2 arcseconds across, and details on any smaller scale are blurred out. The Space Telescope will be able to resolve details right down to the theoretical limit, set by the size of the main mirror relative to the wavelength. At the standard wavelength of 633 nanometres, 70 per cent of the light is focused into a circle only 0.15 arcsecond in diameter, a factor of 10 better than traditional telescopes.

The Space Telescope is also free from stray light from the atmosphere. It will not take in any stray light, for example, from street-lights or office buildings. And it will also avoid the airglow, natural light due to the recombination of ions high in the atmosphere. The net effect will be that the Space Telescope can detect objects 50 times fainter than current ground-based telescopes.

The telescope has a main mirror 2.4 metres in diameter and 0.3 metre thick. It consists of two face plates separated by a honeycomb structure: at three-quarters of a tonne, it is only one-fifth the weight of a solid glass mirror with the same dimensions.

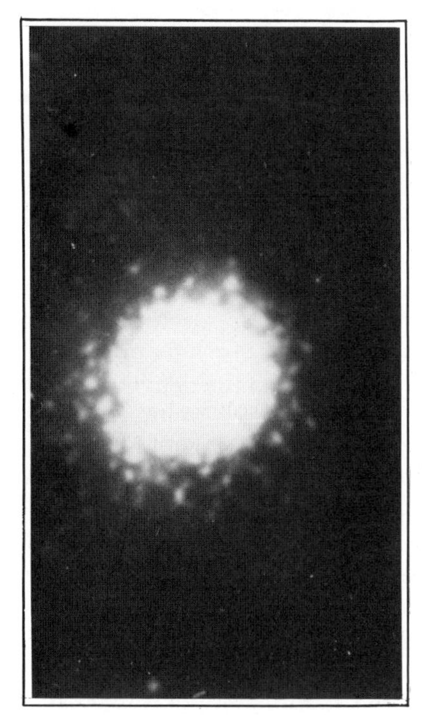

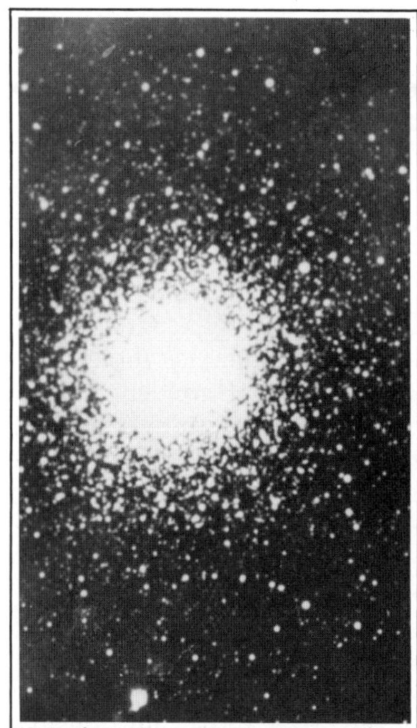

A major advantage of the Space Telescope's position above our blurring atmosphere will be its ability to "see" much finer detail. Its resolution will be 10 times better than any Earth-based optical telescope. These two views of a globular star cluster simulate the dramatic improvement in resolution. A distant globular cluster – perhaps attached to a galaxy in the Virgo cluster of galaxies – appears to present telescopes as a mere fuzzy blur (left), but the Space Telescope will resolve it into its constituent stars (right)

Instruments to detect and analyse the light are housed at the focus behind the main mirror. Five instruments have been chosen to exploit the potential of the telescope: four are American, and one European, to match the 15 per cent of the Space Telescope cost being borne by the European Space Agency.

There are two "cameras", which electronically "photograph" regions of sky. The Wide Field/Planetary Camera will see fairly large fields of view, and so include many objects at once. Fifty interchangeable filters are available to photograph the field in different colours. In the Wide Field mode, the camera covers 160 × 160 arcseconds; in the Planetary mode it is 69 arcseconds square, which will comfortably include all the planets. The European instrument is the Faint Object Camera. To some extent its capabilities overlap with the previous camera, and Space Telescope scientists are anxious that the two should be complementary, rather than compete. The Faint Object Camera is designed to exploit the telescope's potential to the full. It uses a television tube arrangement to count individual photons of light, and reject background noise, so it is extremely efficient and can record the faintest images quickly.

The two spectrographs, however, have very different roles. The Faint Object Spectrograph is designed to obtain spectra of very faint and distant galaxies and quasars. It spreads this dim light only to the extent required to identify spectral lines, and so the detail visible in the spectrum is about one-hundredth to one-thousandth the wavelength in question – but it covers the whole visible and ultraviolet range. The High Resolution Spectrograph looks in detail at brighter sources, at ultraviolet wavelengths which are not visible from ground-based observatories. It can resolve spectral lines as narrow as one-hundred-thousandth of the wavelength observed. The fifth instrument is a photometer, to measure accurately the brightness and polarisation of stars. It can measure changes in brightness as rapid as 60 000 fluctuations per second.

The 10-tonne orbiting observatory has a design life of 15 years, although mission scientists hope it will last for at least 5 years longer. Shuttle astronauts will service the telescope and instruments every 2½ years. If necessary, the shuttle can return the telescope to Earth for repairs.

The Space Telescope is a joint project, with the American space agency NASA contributing 85 per cent of its costs, and the European Space Agency (ESA) giving 15 per cent. Europe's centre for coordinating observations from the Space Telescope will be at the headquarters of the European Southern Observatory, at Garching near Munich. The European Southern Observatory is a consortium of six nations (France, West Germany, Holland, Belgium, Sweden and Denmark) formed to study the southern skies, which are not visible from Europe. It was founded in 1962, and now has a dozen telescopes at its observatory at La Silla in Chile, one of them a 3.6-metre reflector which is the world's ninth largest telescope. It was initially administered from Hamburg. After running into technical and administrative problems, CERN (the European Organisation for Nuclear Research) at Geneva took it under its wing in 1970. Five years later the West German government offered the observatory its present administrative site at Garching.

The "brain" controlling the Space Telescope's "eye", however, will be the Space Telescope Science Institute, at Johns Hopkins University at Baltimore in Maryland. The institute will assess proposals put to it by different groups of astronomers, decide on an observing schedule and send the schedule to the nerve centre of NASA's satellite control network, the Goddard Space Flight Center at Greenbelt, also in Maryland.

At Goddard, a team from the institute – the Science Support Center – will be working with NASA engineers on the practical aspects of controlling the Space Telescope – for example, ensuring that the satellite is correctly oriented so as to point its solar power panels at the Sun while it is observing. The Science Support Center at Goddard will meld these requirements with the scientific schedule from the institute, which will list the objects to be observed, how long the telescope needs to look at each, and which of the five detectors it will use. NASA will send the resulting instructions up to the Space Telescope every day.

The Space Telescope will be in such a low orbit that it will usually be out of sight of NASA's worldwide network of radio dishes which communicate with satellites and space probes. Goddard will therefore communicate with the Space Telescope via intermediate satellites in much higher orbits. NASA will establish this set of Tracking and Data Relay Satellites in 1983 for communication with the space shuttle, and especially with Spacelab, which will need to send information back to Earth quickly to avoid a lot of data storage in the lab itself. Each of these satellites can "see" almost a hemisphere of the Earth, and one or another will always have the Space Telescope in view.

The Space Telescope will make important new observations of all types of astronomical objects, from the planets in the solar system to galaxies forming over 10 000 million light years away. One of its most vital tasks is to refine definitively the scale of astronomical distances. In this step-like progression, the first step can be put on firmer ground by accurate measurements of the movements of nearby stars (parallax) and star clusters (moving cluster method).

But the main controversy currently centres on distance to galaxies beyond the Andromeda Galaxy. The most reliable distance indicators for other galaxies are the Cepheid variable stars, at present not detectable in galaxies farther than Andromeda. The Space Telescope will be able to see Cepheids in the galaxies of the nearest large cluster, the Virgo cluster of galaxies, and settle the extragalactic distance scale once and for all. Indirectly, this will also

Artist's impression of the Space Telescope in orbit. Due to be launched by the space shuttle in 1986, this 10-tonne observatory, orbiting above Earth's shifting, absorbing atmosphere, will return data from objects far more distant than any that have hitherto been visible from the ground

settle the problem of the age of the Universe, for which current estimates range from 10 to 20 billion years.

The Space Telescope instruments will provide more information on the life-cycles of stars. They can investigate the faint white dwarf stars and neutron stars in our Galaxy, which are the corpses of dead stars, and also the double star systems where the companion star's gases are falling onto one of these compact stars – or even falling into a black hole. When looking at the nearest neighbouring galaxies, the Space Telescope will see ordinary stars, not just the superluminous ones, and can put studies of stars in our Galaxy into their universal context.

Quasars will be another fruitful field of study. They are the active nuclei of galaxies, and the Space Telescope will resolve details of their structures, and investigate more fully their spectra – which may also be affected by the outskirts of galaxies lying between the quasars and us.

Perhaps most important, the Space Telescope's great light grasp means it can look farther out into space than ever before, and see more remote galaxies. Because light travels at a finite speed, this means looking back in time, and seeing galaxies as they were when much younger – perhaps seeing right back to the time when galaxies first formed, only a billion years after the big bang. Galaxy evolution, and particularly galaxy formation, are among the major unsolved problems in astronomy today, and the Space Telescope's penetrating stare is essential to their solution.

Index